Environmental Science and Engineering

Environmental Science

Series Editors

Rod Allan
Ulrich Förstner
Wim Salomons

For further volumes:
http://www.springer.com/series/3234

Ole Bruun · Thorkil Casse
Editors

On the Frontiers of Climate and Environmental Change

Vulnerabilities and Adaptations in Central Vietnam

Springer

Editors
Ole Bruun
Thorkil Casse
Department of Society and Globalization
Roskilde University
Roskilde
Denmark

ISSN 1431-6250
ISBN 978-3-642-35803-6 ISBN 978-3-642-35804-3 (eBook)
DOI 10.1007/978-3-642-35804-3
Springer Heidelberg New York Dordrecht London

Library of Congress Control Number: 2013933015

Printed on acid-free paper

Springer is part of Springer Science+Business Media (www.springer.com)

Acknowledgments

The present study builds mainly on material collected in the Danida funded research project, 'Assessing Effects of and Responses to Climate Change on Environment and Socio-economic Development in Mid-Central Vietnam' (2008–2012). The project was jointly conducted by the Institute of Geography, Vietnam Academy of Science and Technology, and the ISG and ENSPAC Institutes at Roskilde University, Denmark. We are grateful to the Danida Consultative Committee for Development Research and to our partners in Hanoi for providing the economic and practical opportunities for research. We further acknowledge the support from the Department of Society and Globalisation, Roskilde University.

Through surveys, interviews, and daily interaction the common people of villages and towns in Central Vietnam provided a wealth of material for this project. We are very deeply grateful to them for offering their time and hospitality during our lengthy stays. Departments of government and administration at all levels have provided introductions, overviews and statistics, acted as guides, and been available for interviewing throughout the process of data collection, for which we are also very grateful.

We are further indebted to the following people for participating in the project or otherwise contributing to the book: Hai Ha Nguyen, Nhinh Do Thi, Dang Thi Thuy, driver Binh, Bodil Folke Frederiksen, Andrew Crabtree, Paul Austin Stacey, Peter Kragelund, Henrik Secher-Marcussen, Tobias Hagmann, Peter Oksen, Fiona Wilson, Hans van der Broek, and Inge Jensen.

Acknowledgments

Contents

Climate Change, Adaptation and the Environment in Central Vietnam

Ole Bruun and Thorkil Casse

Abstract This introduction locates Central Vietnam in the international debate on climate change in developing countries, particularly in terms of socio-economic impacts. Central Vietnam is considered the most vulnerable region of the country in terms of exposure to the effects of climate change. The prospects of a rise in sea level, threats to agricultural production in coastal zones, and risks of increasing weather variability imply higher frequencies of storms, heavy rains and droughts. In order to determine how people, communities and public authorities adapt to new circumstances however, these overall challenges must be placed in a real-life context. There is a gap in the international climate change debate between, on the one hand, a reliance on technical approaches and overall mechanical modeling to countries and regions, and on the other, the perspectives that derive from local environmental data collection and socio-economic analysis. Complexities increase dramatically when working at the lower and intermediate levels: the observed processes of change are not only ascribable to climate change, but to globalization, policy changes, marketization, general economic development, and large-scale human interventions in the environment. There is an urgent need for integrated approaches, such as the building of environmental management into climate change responses, addressing the total impact of livelihood stresses in social vulnerability perspectives, and ensuring that overall adaptation policies adequately address social justice.

Keywords Socio-economic impact of climate change • Global and local change • Social vulnerability • Adaptation • Modernist development policies • Centralist planning • Conflicts with social and environmental objectives

O. Bruun (✉) · T. Casse
Department of Society and Globalisation, Roskilde University, Universitetsvej 1, 4000 Roskilde, Denmark
e-mail: obruun@ruc.dk

T. Casse
e-mail: casse@ruc.dk

O. Bruun and T. Casse (eds.), *On the Frontiers of Climate and Environmental Change*, Environmental Science and Engineering, DOI: 10.1007/978-3-642-35804-3_1,

1 Challenges

Vietnam unquestionably belongs to a group of development countries with high population densities and sensitive ecosystems, to which climate disruptions pose a huge challenge. Countries like India, Bangladesh, Madagascar, Zimbabwe, Mozambique, Haiti, Vietnam and many others, already suffer the combined effects of adverse conditions like population pressure in coastal areas, environmental degradation, inadequate technologies, underdeveloped economies and the differentiating forces of globalization.

A variety of international sources point out that Vietnam is both highly vulnerable and heavily exposed to the risks of increased weather variability and climate change, including storms, droughts, flooding and saline water intrusion (e.g. Chaudhry and Ruysschaert 2007; ADB 2009; MONRE 2009; WB 2010). Particularly, the socio-economic impacts are a concern to Vietnam's present phase of development (McElwee 2010). Climate change is further believed to threaten the poor and upset poverty reduction (Sperling 2002; MONRE 2007). At the same time Vietnam is already among the major receivers of general aid funding, amounting to 30$ per capita per year over the period 2005–2010 (DAC). A series of typhoons during the last decade, most recently the 2009 typhoon Ketsana, caused enormous damage and intensified a focus on climate change in development aid; following World Bank evaluations and appeals from the Vietnamese government, massive donor funding is now directed towards climate change adaptation and mitigation.

The present study contributes to country-specific and regional analyses of climate change impacts and adaptations. By focusing on Central Vietnam it will place overall challenges in a factual context and show how people, communities and public authorities cope with new circumstances. Noted above, complexities increase dramatically when working at this intermediate level. Processes of change tend to relate only implicitly to climate change, but more explicitly to the overall factors of development. Addressing such interacting processes of change calls for interdisciplinary methodologies and explorative means of interpretation. Indeed, climate change impacts, like disasters, cannot be studied separately from 'normal' life. Changeable as it may be, the risks involved in climate change must be seen against the vulnerability created by many people through their normal existence (Wisner et al. 2004, p. 4). Yet the unprecedented confluence of these powerful and transformational forces break up society along new fault lines, and pushes resource exploitation towards new frontiers.

Picking up this challenge, the present work builds on extensive field studies in Central Vietnam and brings together inputs from a range of disciplines, including economics, human geography, science, political science, anthropology and the humanities. It concentrates on rural areas, where the majority of the population lives and whose livelihoods are predominantly drawn from natural resource management, and it spans all three topographical zones of Central Vietnam (coastal, mid-elevation and highland). It is academic in orientation, but speaks as much to developers, planners and decision makers. The strictly hard science components are limited to overviews of climate and agricultural data, while the bulk of the material attends to socio-economic, human-geographical, political and cultural aspects of vulnerability and adaptation.

2 Global to Local Change

We do not challenge the necessity of climate change action to reduce global warming and its potentially catastrophic consequences (e.g. Giddens 2009). Nevertheless, we find it imperative to raise some critical issues relating to the implications of a narrow climate focus and its devolution to national and local politics, where it inevitably introduces a new terminology, sets new agendas and potentially reshuffles power relations. There is a risk that it may overshadow existing concerns, such as poverty alleviation, good governance and sustainable development. There is a need to recontextualize the climate change rationale into a societal reality.

Global concern over a long-term change in weather patterns has been nourished by rising average temperatures, which are likely to increase the future frequency of extreme weather events such as storms, heavy rainfall and heat waves. It will further imply rising sea levels due to the melting of polar ice sheets, with major long-term impacts on low-lying coastal regions and river deltas, including the most densely populated areas of Vietnam. The country has an extensive coastline, stretching over more than 3,000 km, with crucially important river deltas on the Red River and the Mekong River. Most parts of the country are subjected to tropical cyclones and monsoon rains, at the same time making them heavily exposed to increasing weather variability and climate change. "Climate Change's Impact on Natural Hazards in Quang Nam Province, Mid-Central Vietnam" shows some key data for Central Vietnam and makes a comparison with national climate scenarios, in order to forecast trends in flooding, drought and landslides. For the central region, changes in temperature and precipitation are not significant as yet, while an increasing intensity of typhoons may be documented. Nevertheless, tropical cyclones appear to be more powerful in strength and those at Beaufort wind force 12 or more have increased from two in each of the decades 1981–1990 and 1991–2000 to five in the period of 2001–2008 (Van Tan 2010, p. 35). As shown below, however, other observable changes like increasing flooding disasters and landslides are inherently multi-causal.

In an absolute sense, climate change relates to the statistical distribution of weather patterns over extended periods of time. In common parlance, it is qualified as anthropogenic climate change, also known as global warming or anthropogenic global warming.[1] In a given locality global warming, implying something

[1] The UN Framework Convention on Climate Change thus defines climate change as 'a change of climate which is attributed directly or indirectly to human activity that alters the composition of the global atmosphere and which is in addition to natural climate variability observed over comparable time periods' (UNFCC).

gradual and uniform, tends to be a misnomer, since its impact is both uneven and unpredictable: it may appear as temperature oscillations, increasing fluctuations over the year, erratic patterns, or show up as out-of-the-ordinary natural hazards. Besides, changing climatic conditions are not inherently equivalent to disaster and projected changes in climate may have both beneficial and adverse effects in a given locality, for instance relating to agriculture, ecosystems, water resources and human health. The greater and faster the change in climate conditions, however, the greater likelihood that adverse effects will dominate. "Is Climate Change a Reality for Agriculture in Quang Nam Province?" examines key trends in agricultural production in Central Vietnam, including recent land use changes, new crop varieties, general intensification, and increasing uses of fertilizer and pesticides. Despite the general experience of an increasing frequency of typhoons, heavy rain and flooding, available materials show substantial overall growth of all forms of agricultural production, including grain cropping, aquaculture and forestry. Nevertheless, efficient physical planning and coastal protection remain great challenges for adaptation.

Overall climate change merges with a great amount of local weather variation to create a range of outcomes. Central Vietnam is considered the most vulnerable region in terms of *exposure* to the effects of climate change (WB 2010), including, particularly, storms, flooding, landslides and droughts, and although the region does not overall rank among the most critical in *sensitivity*, particularly poverty and economic diversification are major concerns. Historically, Vietnam's exposure to extreme weather has resulted in countless disasters and loss of lives, but has also provided its inhabitants with a rich experience in coping with the effects of droughts, flooding, storms, and similar events on agriculture and society, both individually and collectively in a strong state: sensitivity to climate change must be seen against the historical and societal context. Hence, an evaluation of the present impact of weather and climate events in Central Vietnam has to take two separate factors into account.

First, this is a historical hazard-prone region, part of the Southeast Asian 'typhoon belt', in which the season from July through November brings frequent heavy winds, on average four to six times per year, maturing into typhoons (tropical cyclones with wind speeds over 32.8 m/s or 118 km/h). There is also plenty of rain that feeds a natural abundance of vegetation and a huge biodiversity. Until quite recently, only a narrow belt of the lowland coastal areas and river deltas were inhabited by sedentary agriculturalists, while inland and highland forest regions were 'shatter zones' (Scott 2009). These were inhabited by slash-and-burn agriculturalists and hunter-gatherers of a great ethnic variety. The historical context and present situation of one such ethnic minority is described in "Interacting Cultural and Environmental Change: The Co (Cua) Minority of Central Vietnam", showing that the simultaneous felling of the forest and moving of the population in ambitious state schemes has altered both landscapes and habitation patterns.

Since the earliest historical records, Vietnamese culture has embraced the environment in transformational schemes. As an 'ancient hydraulic civilization' it has harnessed rivers and constructed dikes and irrigation canals for the cultivation of lowland paddy rice while restraining the natural occurrences of flooding in gradually increasing areas, inevitably taking risks in the process. The great spatial variety of Vietnam provides very diverse natural and man-made environments, and livelihoods are often earned in locations where opportunities and hazards coalesce.

When man-made dikes burst during heavy rains or when flooding reaches higher-than-usual levels, natural hazards may turn into societal disasters. There is a long historical record of flooding and disaster: the city of Hanoi has recorded 20 'historical floods' during its 1,000 year history as a capital, in the last century in 1915 and again in 1971, when a '250 year flood' overwhelmed the dikes and according to some estimates killed as many as 100,000 people. Typhoons are a perennial threat; the Haiphong typhoon of 1881 was among the deadliest in human history, claiming an estimated 300,000 lives. Most recently, 1997 typhoon Linda killed 3,000 people in the Red River Delta, and floods in 1999, 2008 and 2010 also claimed many casualties across Vietnam. As a perennial threat to Vietnam as a whole it applies to Central Vietnam more specifically. Recent historical records include a terrible storm that swept away a French-built railway line between Danang and Hoian in 1915/16, the highest ever recorded flooding in 1964 (during the most active typhoon season ever recorded), a great flood in 1999 claiming 700 lives, and serious floods in 2007 and 2009. To these events may be added the memories of local people, in which the endless loss of crops, domestic animals, houses and property in seasonal, but unpredictable floods, and the destruction of houses during typhoons, are closely interwoven with family history. There is a constant and inescapable element of risk in all human and agricultural activity in this historical natural hazard-prone region, but also a long historical experience in coping with changeable conditions.

Second, it must be stressed with even greater weight that external, anthropogenic climate changes, whatever their scope, invariably interact with a range of local environmental interventions. These are occurring at an accelerating rate in the present phase of Vietnam's social and economic development. More specifically, industrialization, infrastructure and other construction (see "Paradoxes in Adaptation: Economic Growth and Socio-Economic Differentiation. A Case Study of Mid-Central Vietnam"), fish and shrimp farming (see "Is Climate Change a Reality for Agriculture in Quang Nam Province?"), dam and reservoir building for hydropower (see "Paradoxes in Adaptation: Economic Growth and Socio-Economic Differentiation. A Case Study of Mid-Central Vietnam" and "Rural Households: Socio-Economic Characteristics, Community Organizing and Adaptation Abilities)", forest clearing for construction and population movements (see "Natural Resource Management Impact on Vulnerability in Relation to Climate Change: A Case in a Micro-Scale Vietnamese Context" and "Interacting Cultural and Environmental Change: The Co (Cua) Minority of Central Vietnam)", and plantation forestry development (see "The Push for Plantations: Drivers, Rationales and Social Vulnerability in Quang Nam Province, Vietnam") lend a whole new set of circumstances for all land and resource uses in Central Vietnam. Similar transformation processes can be found in other regions of Vietnam, such as the Mekong Delta (Fabres 2011). Indeed, both historically and in modern times the Vietnamese have engaged with their physical environment in a transformational fashion and pushed for an expansion of resource uses as a basis for a growing population. The sense of urgency in integrating environmental management with climate change response is further discussed below.

Climate change data may assume absolute value over time, but as evident from the above, climate change impacts and adaptations hardly constitute a meaningful field of study in isolation: there is no clear baseline or status quo to measure against, neither historically, nor today. Presently, climate data may only be applicable at the community level in a very narrow sense. For instance, rainfall and its temporal distribution may be measured at local stations. But for the farmers in much of Central Vietnam, water resources are drawn from rivers, streams and canals, the levels and flows of which depend on a range of factors, including upstream forest cover, hydropower construction, major reservoirs, local dikes and general water management. Similarly, crop growth and biomass production may be measured separately as a function of temperatures and precipitation, but the individual farmer participates in a constant process of grain hybridization, improved fertilization and pest control, and increasing global market integration, often leading to entirely new cropping or land use patterns (see "Paradoxes in Adaptation: Economic Growth and Socio-Economic Differentiation. A Case Study of Mid-Central Vietnam"–"The Push for Plantations: Drivers, Rationales and Social Vulnerability in Quang Nam Province, Vietnam"). Compared again with the Mekong Delta, the effects of climate change on rice production may be a reduction of the arable area, which is responded to by the increasing use of agrochemicals which again will spell other long-term challenges (Sebesvari et al. 2011, see also "Vietnam's Food Security: A Castle of Cards in the Winds of Climate Change"). Inevitably, the impacts of local climate disruptions merge with impacts of changing lands uses as well as environmental disruptions and interventions to form multiple, interacting processes of change.

3 Key Livelihood and Climate Problems

With Vietnam's implementation of Doi Moi (renovation) policy in 1986, initiating a 'socialist-oriented market economy', the Communist organization of production in reality ended (see "Paradoxes in Adaptation: Economic Growth and Socio-Economic Differentiation. A Case Study of Mid-Central Vietnam"). The country entered a steady transformation process towards market-driven forms of production, while upholding one-party rule. With accelerated transformation and land reform, the rural population gained new opportunities, but was at the same time faced with a range of new stresses, such as insufficient land, water and capital, eroded security, and a range of new inequalities relating to income, land and labour. Most critically, a rising differentiation between coastal and inland districts and between rich and poor generally impact adaptation capabilities negatively. Both opportunities and stresses are increasingly individualized in rural communities, the texture of which begin to break up from the impacts of marketization, private business, land transfer, wage labour and migration.

Albeit trivial, but substantiated by a large interdisciplinary field material, we argue that for rural people climate change must be seen in the context of those other stress factors already present. In Central Vietnam as elsewhere, common stress factors in rural areas relate to the conventional challenges of resource exploitation, demography and weather events in agricultural societies, including land and water scarcity and inescapable poverty. In simple terms, but nonetheless highly significant, farmer households in climate sensitive areas are less burdened by the individual components of their livelihood stress than by their total impact. Conversely, improved livelihoods in general will better enable them to deal with any single factor separately, including that of climate change.

Despite possible climate change impacts, Vietnam in general and the Central region has experienced rapid economic growth and increasing export earnings (see "Paradoxes in Adaptation: Economic Growth and Socio-Economic Differentiation. A Case Study of Mid-Central Vietnam" and "Is Climate Change a Reality for Agriculture in Quang Nam Province?"). GDP per capita has increased rapidly over the last 15 years, and the positive economic growth has improved livelihoods and reduced rural poverty levels. However, what could be interpreted as signs of decreasing vulnerability, due to economic growth and diversification in times of high demand for wood products, shrimps, rice and so forth, might turn out to increase vulnerability in times of substantial climate changes. Local people might adapt to weather events by cutting the rotation period for acacia plantation forestry, constructing houses with a first floor to rescue human and valuables, in the event of rising water levels or by purchasing boats. In this sense, the people of Quang Nam already adapt to circumstances. The problem is that these new economic activities are highly vulnerable to typhoons and flooding disasters. Only when disasters happen years apart and the economic gains from consecutive regular harvests of timber or shrimps can balance out eventual losses, will these adaptation strategies appear feasible. Our research tends to indicate that the poorest local people pay the price for the increased frequency of disasters, due, in part, to their dependency on casual work and small incomes from acacia or shrimp farms.

Thus, in the short term, hard-won gains are hardly threatened by climate change, but rather by an increasing social differentiation and the rising costs of living and of public goods. With a progressing liberalization of land tenure, capital accumulation opportunities in rural areas have increased and given rise to intense competition over land and resources. As analyzed in "The Push for Plantations: Drivers, Rationales and Social Vulnerability in Quang Nam Province, Vietnam", many farmers have turned to plantation forestry on gradually increasing plots as an effect of both government programs and market forces. Villages are gradually transformed in the process, since rich and middle-income households as well as outside investors can afford the necessary investments, while poorer households may be marginalized. Pro-poor growth tends to be relegated in the process and social conflict may increase in the old villages. "Rural Households: Socio-Economic Characteristics, Community Organizing and Adaptation Abilities", drawing on the joint socio-economic survey (described below in the methodology section) and fieldwork in rural areas shows how diverse groups of households cope differently and have varying levels of vulnerability. A great many households are unable to cope with the tough new market conditions, such as due to disabilities, disease, old age, too many children or other adverse conditions. "Health Impacts of Climate and Environmental Change: Awareness and Challenges to Adaptation" takes a closer look at the public health aspects of present climate and environmental change, and discusses the present health situation in selected rural areas. Despite gains in terms of a decreasing occurrence of infectious diseases, many rural households report an increase in health problems.

Economic growth has facilitated a downward trend in natural hazard fatalities, but at the same time has caused a growing trend in economic losses. The socio-economic survey questionnaire also examined to what extent vulnerability can be monitored in quantitative and economic terms. The analysis, presented in "Livelihood Stresses Under the Constraints of Climate Change Vulnerability in Quang Nam", shows the case of the Ketsana typhoon in 2009: apart from the poor households at the lowest end, the households most seriously affected were those having invested in acacia forestry. A low degree of institutional adaptation means that although disaster-affected households may receive some measure of compensation, disaster relief mostly secures a bare minimum of existence.

As a result of better infrastructure and disaster preparedness, Vietnam now seems less vulnerable compared to many other developing countries. In economic terms the volume of activity and degree of economic diversification in rural areas are key issues for poverty: overall economic diversity is now increasing, while crop diversity tends to be declining in rural areas (e.g. McElwee 2010). These new patterns inevitably contribute to shaping the impacts of natural hazards. Economic losses incurred by extreme weather events are not limited to coastal areas, but now tend to be evenly distributed across geographical zones. While poorer households are more subject to losses from damage on houses, thus being materially vulnerable, higher income groups suffer greater losses in production assets (plantation forestry and aquaculture), thus primarily being vulnerable in a financial sense. Yet the economic diversification of a given area tends to be a moderating factor for all.

Variations between districts of a province seem to be repeated in variation between provinces. Natural disaster vulnerability varies significantly at the provincial level and the present study may indicate that (by multivariable linear regressions), substantial parts of this can be explained by differences in socio-economic development. As described in "Climate Change's Impact on Natural Hazards in Quang Nam Province, Mid-Central Vietnam", higher provincial per capita income appears to reduce provincial natural disaster fatality rates, while inequality, poverty, infant mortality and urbanization were significantly and positively correlated with fatalities. This is consistent with our general argument, that the key concern should be the aggregate impact of stress factors. It remains an issue, that despite a declining rate of disaster fatalities, large groups of households are highly vulnerable when disasters strike and are left with either a minimum amount of state relief, or food handouts and unsystematic support from NGOs.

4 Government and Institutions Matter

Both climate and environmental change, appearing to interact in complex patterns at the local level, call for adaptive forms of government. When impacts overwhelm local adaptive capacity, local communities need external support in their adaptation efforts. A prerequisite for strong local governance institutions is, of course, an efficient national government that can devolve the necessary power down through the system; an issue for national-level government institutions is therefore to identify the most effective means to support local adaptation. Actual adaptive capacity depends heavily on local dynamics, however, in which the livelihood impacts for vulnerable people and local governance structures at district level are in focus. Overall, a crucial parameter for success is the extent to which government policies can promote a synergistic relationship with local government institutions and vulnerable groups.

Differing constellations of actors and policies are essential at different stages of climate, weather or environmental disruption impacts. Early warning systems and training are crucial elements of enhancing livelihood resilience, whereas financial capital, including access to credit and insurance, are vital for long-term recovery. The policy challenge is to identify specific needs in response to given circumstances; that is, which assets will provide the greatest resilience and adaptation capacity. Yet the greatest adaptation capacity of communities will be reached when policy and finance are accurately aimed and efficiently and accountably applied. It is definitely an asset in Vietnam that a reasonable level of governance and public services are available in rural areas, and coordinated response is achievable. However, balancing government inputs with the adaptive capacities in local communities necessitates the identification of relevant local partners, including community and civil society organizations. Presently, as seen in "Impediments to Climate-Induced Disaster Management: Evidence from Quang Nam, Central Vietnam" and "Rural Households: Socio-Economic Characteristics, Community Organizing and Adaptation Abilities", the old communist mass organizations have a near-monopoly on rural organization, and the self-organizing potential of local and non-government associations is poorly utilized.

Immediately after the war, the Socialist Republic of Vietnam employed population, land use and forestry practices expressive of a high-modernist thinking, with serious repercussions for the environment, such as a rapidly declining forest cover. In the later phase, particularly forest policies have been revised and forest cover has increased by means of mainly plantation forestry. Many aspects of the development process are still subjected to centralized state planning, which raises many issues of rights, participation and sustainability. Several aspects of how the state 'sees' development, such as favoring large-scale social and technical engineering, employing high-modernist ideology, upholding centralized Party-state leadership, and keeping a level social terrain by means of an incapacitated civil society (Scott 2004, p. 5), are still oriented towards an optimistic and uncritical mastery of both nature and society that does not always fit in easily with global agendas. Furthermore, social science studies of climate change increasingly focus on the mechanisms by which climate change discourses and technocratic, large scale mitigation and adaptation interventions inadvertently produce new risks, potentially eroding livelihoods and alienating affected groups (e.g. Marino and Ribot 2012).

Public information dissemination and awareness-raising are important elements in climate and environmental change, such as described in "Climate, Environment, and the Role of Media". Equally relevant is the extent to which alternative visions and civil society voices have access to conventional and new digital media. In Vietnam the media in principle are controlled by public authorities and are intended to support state policy. But investigative journalism is on the rise and the media are increasingly used to express discontent with land grabs, unequal development, pollution, illegal mining, environmental problems and so forth.

Everywhere, government may be expected to provide for the health, safety and welfare of local communities. Worst-case scenarios emerge when the government fails to protect the poor and the environment, and the local population in response rejects their own responsibility to change unsustainable land management arrangements. Thus social contracts break down. For instance, in a study from northern Vietnam, the majority of farmers attributed the responsibility for flooding to the mismanagement of the artificial hydropower reservoirs and denied that they could adopt measures to ease the detrimental effects of flooding (Schad et al. 2012). Correspondingly, our own study shows that at times, centralist planning antagonizes farmers, such as when hydropower construction increases flooding levels in downstream communities, or when population movements affect old communities and ethnic minorities (see "Paradoxes in adaptation: Economic growth and socio-economic differentiation. A case study of mid-central Vietnam", "Rural Households: Socio-Economic Characteristics, Community Organizing and Adaptation Abilities", "Natural Resource Management Impact on Vulnerability in Relation to Climate Change. A Case in Micro-Scale Vietnamese Context" and "Interaction Cultural and Environmental Change: The Co (Cua) Minority of Central Vietnam"). The continuance of aggressive and modernist development policies, as ingrained in industrial agriculture with heavy chemical inputs, may further threaten food security in rural areas in times of crisis (see "Vietnam's Food Security: A Castle of Cards in the Winds of Climate Change").

5 The Scope of the Study: Objectives, Concepts and Orientation

A series of key concepts, interrelated in scope and orientation, form the joint background of the study. Some initial delineation is presented here, leaving the qualification and application to individual chapter authors. Our common perspective makes a basic distinction between *natural hazards*, including those deriving from climate variability, and people's *vulnerability* to hazards. This implies that social structures in the general sense are the cause of vulnerability as much as they may respond to external conditions (e.g. Hewitt 1983; Adger 1998; Wisner et. al. 2004).

Social vulnerability, understood as an inadequate capacity of individuals or groups to cope with and recover from the impact of hazards, is therefore a characteristic of society and its underlying social, economic and political conditions (e.g. Bankoff 2003; Gaillard et. al. 2007; Schipper 2007).[2] Social vulnerability deals with the causes and distribution of vulnerability independent of, or before the event of a natural hazard. For instance, this is captured in the disaster 'pressure and release' (PAR) model (Wisner et al. 2004). This traces the progression of vulnerability through root causes in society itself, dynamic pressures in institutions and macro-conditions, and unsafe conditions of all sorts in local communities; disasters then occur as the culmination of progressing vulnerability on one side and natural hazards on the other.

Since the social dimension of vulnerability spans a range of entitlements and institutions, local efforts towards improved livelihoods inevitably depend on conditions in the broader society. Vulnerability and poverty tend to be closely related, and measures to alleviate them must correlate. An important observation is that 'although poor people are usually among the most vulnerable, not all vulnerable people are poor' (Moser 1998, p. 3). This means that in contrast to poverty as a simple expression of income level and well-being, vulnerability must include the risk of a household becoming poor (Chaudhuri et al. 2002). In tangible terms, the ability to cope with the impacts of natural disasters is a function of various socio-economic factors, including age, gender, land property, capital, education and access to financial transfers from outside. Deficiencies and weaknesses among these factors are similar to those resulting in poverty.

Further highlighting relativity, the presence of inequality can augment the collective vulnerability of a community. Noted above, in quantitative terms vulnerability may be measured in the losses incurred by the households and the disaster relief through the institutional framework, but clearly both the institutional capacity and aspects of poverty and inequality require qualitative and contextual assessments. For institutional adaptation, there is a crucial distinction between starting-point and end-point vulnerability (O'Brien et al. 2004). End-point vulnerability is an outcome of vulnerability in linear terms, linking climate change projections to an impact on an exposure unit (for example a household). Starting-point vulnerability, on the other hand, is contextual, based on a non-linear process of a multidimensional view of climate change. It has been pointed out that up until now, the end-point or outcome vulnerability has dominated the debate, and adaptation policies are defined quite narrowly to that end. Another related approach compares first generation (more technically oriented) and second generation (more society oriented) vulnerability assessments (Füssel and Klein 2006). In the second-generation assessments, non-climatic drivers (demographic, social, economic and political) are considered to affect the sensitivity of a community or a society

[2] Obviously, a social vulnerability perspective has a different orientation than a technical vulnerability perspective (such as applied by the IPCC), which in general terms defines vulnerability as a function of exposure, sensitivity and adaptability.

to respond to climate changes. Economic globalization and urbanization represent examples of such non-climatic drivers.

Thus, the political and socio-economic processes of marginalization and differentiation are crucial for understanding and assessing vulnerability, and they surface in very tangible factors like poverty and low levels of resource entitlements and economic opportunities (e.g. Brooks 2003; O'Brien et al. 2004; Marino and Ribot 2012). Particularly relevant for the political context of Vietnam, as a one-party state, is that social and political inclusion or exclusion is crucial for vulnerability, since marginalized groups may lack voice and influence on decision making (Eriksen et al. 2011; Bruun 2012). Approaches to vulnerability based on *entitlements*, understood as resources available to households based on their common assets, production and institutional arrangements, thus pursue the socially generated impediments to disaster readiness and management and to adaptation (see "Impediments to Climate-Induced Disaster Management: Evidence from Quang Nam, Central Vietnam" and "Natural Resource Management Impact on Vulnerability in Relation to Climate Change. A Case in Micro-Scale Vietnamese Context").

Vulnerability may also be monitored in quantitative terms, such as by means of a three-tier framework consisting of poverty, inequality, and institutional adaptation (Adger and Kelly 1999). Noted above, when looking at the typhoon Ketsana in 2009, the poorest households as well as those engaged in acacia production and shrimp farming were hardest hit (see "Livelihood Stresses Under the Constraints of Climate Change Vulnerability in Quang Nam" and "Natural Resource Management Impact on Vulnerability in Relation to Climate Change. A Case in Micro-Scale Vietnamese Context"). A declining diversity of crops and of household livelihoods in many rural areas coalesces with eroding social safety nets in the reforming society: they may eventually clash with the likely impacts from climate change (McElwee 2010, p. xiii). In particular, a livelihood strategy based on acacia mono-cropping presents a double-edged sword. On the one hand, the acacia tree species is efficient in reducing water flow-off and economically profitable, thereby reducing social vulnerability and leading to a general increase in farmer incomes. On the other hand, dependency on monocropping exacerbates the impact of natural hazards like typhoons and droughts, thus potentially differentiating households between those with greater assets and those with no alternative income sources to turn to. Quantitative approaches to vulnerability may further elucidate the stratifying mechanism in climate interventions (Marino and Ribot 2012), such as when the poor and landless are increasingly pushed towards casual work on wealthier households' plantations.

The above is in no way a denial of the significance of geographical location in the sense of proximity to the source of a natural threat or high exposure to such (e.g. Cutter et al. 2003, p. 4). *Physical vulnerability* associated with a given area must be taken into account in any community assessment and comparison (e.g. Adger 2006), and the geographical characteristics of local communities, such as land and forest entitlements, water availability, distance to markets and accessibility by road are crucial denominators (see "Paradoxes in Adaptation: Economic

Growth and Socio-Economic Differentiation. A Case Study of Mid-Central Vietnam", "Impediments to Climate-Induced Disaster Management: Evidence from Quang Nam, Central Vietnam", "Livelihood Stresses Under the Constraints of Climate Change Vulnerability in Quang Nam" and "Rural Households: Socio-Economic Characteristics, Community Organizing and Adaptation Abilities"). Yet location in itself is not a consistent factor (for disaster losses and economic vulnerability across geographical zones, see "Livelihood Stresses Under the Constraints of Climate Change Vulnerability in Quang Nam"). Vulnerable people are also found outside the most vulnerable places, and even fragile physical environments may provide access to wealth for some people (Moser 1998, p. 3; Chaudhuri et al. 2002; O'Brien and Leichenko 2000, p. 224). Highly risk-prone areas in Central Vietnam offer great economic opportunities, while the burden of poverty has been shifting across geographical space.

Place is itself subject to change: with the growing impact of *environmental interventions* (understood as alterations in the physical environment for economic or demographic purposes), including both mitigation projects targeted at reducing risks and those deriving from economic development as described above, place-bound vulnerabilities will constantly alter for better or worse. When seen in even a brief historical context, the ethnic mountain dwellers of Central Vietnam, who up until the war period controlled huge natural resources, have backslid to become without comparison the poorest populations segment in present society (see "Interacting Cultural and Environmental Change: The Co (Cua) Minority of Central Vietnam"). Poor fishing communities struggling with declining marine resources have suddenly found their coastal setting a new spot for holiday resort development, providing new land and labour opportunities. Notwithstanding structural conditions, both social and place-bound vulnerabilities are increasingly dynamic factors in industrializing and globalizing locales. It may further be argued with regard to physical vulnerability, that any complex society can be expected to offer some measure of protection against and compensation for place-specific exposure, since local governments in principle provide for the health, safety and welfare of their local communities.

In a Vietnamese context the *household* (a cohabitation unit usually spanning several generations and primarily consisting of members related through marriage and blood) still tends to be the basic socio-economic unit. Household vulnerability strikes through in key variables, such as access to land, livelihoods and employment, household composition, education, social group belonging, ethnicity, and access to financial and other transfers from outside (the special relationship between vulnerability and poverty will be discussed in "Livelihood Stresses Under the Constraints of Climate Change Vulnerability in Quang Nam" and "Rural Households: Socio-Economic Characteristics, Community Organizing and Adaptation Abilities"). Living with uncertainty and climate variability are constitutive elements of small farmer production in most parts of the world. Efforts to cope with stresses and external shocks are inherent in local production systems and household strategies. When seen in the perspective of the rural household, climate change represents a *stress factor* similar to other stress factors already present, such as those deriving from the interacting processes of change described above. Poor or vulnerable households frequently suffer multiple and mutually reinforcing shocks from very diverse circumstances. These include living on marginal and risky land, lacking capital, having poor housing, being poorly educated and being prone to suffering health problems (see "Climate, environment, and the role of media"). Simply put, the poor suffer not only from the event of a given disaster, but as much from the social conditions under which they already live (e.g. Adger and Kelly 1999).

Adaptation is commonly defined as adjustment in natural or human systems in response to actual or expected climate stimuli and their effects, thus carried out either in direct response to or in anticipation of changing climatic conditions. In the specific context of socio-economic analysis, however, it may be understood as combined decision-making processes and actions, undertaken to maintain capacity in the face of current and predicted environmental change (e.g. Nelson et al. 2007). It can therefore activate a range of indigenous knowledge backgrounds and informal institutions contained in local culture as well as rely on the formal institutions of state and government. These include both conventional and targeted climate change and disaster prevention institutions (see "Impediments to Climate-Induced Disaster Management: Evidence from Quang Nam, Central Vietnam"). *Adaptive capacities* by which to adjust and adapt, are inherent in any natural or social 'system'. However, in a social context they are inevitably subject to questions of Who? and How?, accentuating issues of poverty, social differentiation, migration, human rights and democracy.

Obviously a great breath of studies is needed to get a clear sense of the multiple adaptations that individual households and communities perform. However, there is a lack of consensus about which factors determinate adaptive capacity and how to measure success; some scholars indicate the possibility of ending up with an adaptive capacity deficit (Williamson et al. 2012). If disaster plans prove unrealistic in the advent of natural hazards, such as when disaster relief does not reach out (see "Impediments to Climate-Induced Disaster Management: Evidence from Quang Nam, Central Vietnam" and "Livelihood Stresses Under the Constraints of Climate Change Vulnerability in Quang Nam"), adaptive capacity in obviously at fault. But there is also a crucial time dimension. This is because adaptation takes place in response to events or a series of events over time, which demands a period of consideration that tends to delay action. Thus, the full sequence of climate events, change of strategies, and real-life adaptations only becomes visible over a temporal length. Studying the process demands repeated data collection over a lengthy period of time. And making sense of our field experiences as researchers further expands the time frame, to the point where the 'adaptations' documented may no longer be sufficient or even relevant. As inferred above, the range of qualities enjoyed by the broader society also determine the space for opportunity.

The above conceptual discussion focuses on the consecutive levels of land uses, resource management, institutions and governance. As indicated in the discussion of physical vulnerability, however, individual actions and social processes cannot be viewed in isolation from the natural resource base with which they interacting. Placing them in a joint socio-ecological framework is tempting (Berkes and Folke 1998; Adger 2006), but fosters other challenges. For example, studies of 'environmental stress factors' commonly involve both 'social and ecological systems' (Berkes et al. 2003) and point to a complexity that may only be investigated through meta-level approaches and broad conceptual constructs. Commonplace terminology in climate change and vulnerability studies, notably concepts of 'systems', 'adaptation', 'capacity' and 'resilience', project images that rural communities are oriented towards harmony and equilibrium and strive to revert to a given and static state. Over the years, successive studies have pointed out that 'natural systems' may not have a single steady state, but organize around a series of possible states (e.g. Holling 1973; Nelson et al. 2007), or may be characterized by non-equilibrium dynamics (e.g. Folke 2006). This lends further credibility to complexity theory which rules out certainty in principle (e.g. Prigogine 1997; Mitchel 2009). Speaking of 'social systems' in local communities as the other side of the equation, and even seeing social structures as integral to nature (Berkes and Folke 1998; Adger 2006), further adds to the vagueness in integrated socio-ecological approaches. Social 'systems' easily become corrupted and inadvertently geared towards short-term gains while degrading the natural environmental.

With the present rate of development in rural areas of Central Vietnam we continuously see land use forms discontinued and breaking up as both natural and social 'systems'. This frequently results in villages communities leaping into a whole new state. Examples include labour intensive grain and vegetable cropping for subsistence converted into extensive cash cropping due to labour migration, rice fields being pooled into large fish and shrimp farms, hilly forest land being cleared for acacia plantation forestry, local population being shifted out for reservoir construction, or entire agricultural areas transforming into industrialized town life. Moreover, and contrary to popular perceptions of rural communities, the present phase of development may be characterized with social and cultural values becoming decreasingly stability orientated, and increasingly geared towards optimally exploiting natural and social systems for a transfer into modern-sector, industrialized, or urban life.

In these circumstances, what does 'adaptation' really imply? Obviously, not every adaptation is a good one (Eriksen et al. 2011), and local 'communities' may in fact desire land use transitions that seriously burden the environment (for the concept and meaning of 'community' in Central Vietnam, see "Livelihood Stresses Under the Constraints of Climate Change Vulnerability in Quang Nam" and "Rural Households: Socio-Economic Characteristics, Community Organizing and Adaptation Abilities").

6 Integrating Climate and Environment

A rapidly growing body of international literature is establishing climate change vulnerability and adaptation as a separate field. It has been bestowed with its own distinct discourse, academic assemblies and publications, and donor organizations have built up targeted aid programs in its wake. There is a risk of separating the 'climate' from the natural environment in which people live, work and sense the world, thereby overexposing technical and meteorological parameters of change. This can also neglect those social, national and global structures which commonly impact the physical environment in developing countries on a far greater scale. And little is gained if climate change response is sidelining the vital areas of accountable governance and sustainable development. Fortunately, a sense of urgency and responsibility feeds the debate on climate change mitigation and adaptation processes. But we wish to emphasize that when global policies and initiatives enter national politics in weak or non-democratic countries, they easily get caught up in the kind of rhetoric that props up regime legitimacy while referring social and environmental ills to incontrollable or outside forces. In the case of Vietnam, central control and political interest in the climate change discourse merge together with overall political priorities, which in practice affects a separation of climate change response from environmental protection (Fortier 2010; Bruun 2012, see also "Natural Resource Management Impact on Vulnerability in Relation to Climate Change. A Case in Micro-Scale Vietnamese Context" and "Interaction Cultural and Environmental Change: The Co (Cua) Minority of Central Vietnam").

The integration of climate change adaptation and mitigation with environmental protection, as much as with any development initiative, is given in a 2006 OECD Declaration and specified in OECD policy guidance (OECD 2009), and has been further developed in many individual donor strategies. Such integration is echoed in the Vietnamese Action Plan for Climate Change (MONRE 2007) and remains a stated goal of the foreign donors contributing to its funding. Yet there are multiple obstacles for carrying it out in practice. These include the procedural and economic priorities laid out in overall government policy, but also cultural issues, power politics and practical set-ups on the ground. After extensive field studies in Central Vietnam we wish to point out that environmental protection and commonplace attention given to sustainable development, are taking the back seat everywhere in favour of immediate economic gains. Most disturbingly, the very conception of environmental protection may gradually be sidelined by the climate change discourse. There are pertinent arguments for further strengthening the environmental dimension of any action on climate change. We propose three such leading arguments.

First of all, there are the technical-administrative concerns. There is presently a risk of both policy making and implementation being compartmentalized into 'conventional' environmental and 'new' climate change organs, each with a narrow accountability towards their respective ministries, policies and donors. New agencies are set up, leading to an incessant flow of policy and strategy papers, and individual ministries have joined in with their own climate change programs. Experience from Central Vietnam calls into question a constricting approach to climate change impacts. Depending on context, it is pertinent to address short-term strategies in exploitation of natural resources, resource degradation, polluting and destructive activities, and irregularities in local land management. All of these are likely to interact with and even aggravate climate change impacts. We believe that in order to make relevant and realistic recommendations for local communities aimed at long-term benefits for their climate change adaptation and resilience, environmental and human land use factors must be taken into account throughout. As noted above, the hydropower construction and changing forestry patterns presently taking place in several provinces of Central Vietnam, appear to impact river flows far more dramatically than climate change. For instance, farmers and city dwellers in Quang Nam experienced unusually rapid and destructive flooding in the aftermath of the 2009 Ketsana typhoon. However, later news revealed that major hydropower plants at the same time released surplus water into the rivers (LAV 2009a, b). The regulatory function of their reservoirs is increasingly called into question, and they may in actual practice amplify natural cycles by retaining water in the dry season and releasing it during rains in order to optimize production. The joint party-state-private push for the expansion of hydropower capacity is enormous and environmental impact assessments (EIAs) are not allowed to delay the process. In fact, evidence suggests that construction work may be started and land use permits issued to developers even before data collection for EIAs is undertaken (Dao 2010).

Implementing first-order adaptation measures is no guarantee that the entire adaptation process is successful (Birkman 2011). As a relevant example, building dykes in the northern Mekong delta enabled farmers to continue their agricultural production, but the adaptation measures had negative effects on the fish stock and increased the dependence of local farmers on agrochemicals (reduction of flood induced fertilization). In another example, it was suggested that the Vietnamese coast line should be divided into risk zones, and to adopt non-infrastructural, low-tech solutions in low-lying zones, such as building houses on stilts and constructing high-lying storm shelters (Boateng 2012). In Quang Nam, rich and middle-income families are now also constructing two-storey houses to shelter food and people during periods of extreme flooding (see "Rural Households: Socio-Economic Characteristics, Community Organizing and Adaptation Abilities").

Second, there are economic considerations. Economic losses from disasters are increasing on a global scale as a consequence of economic development, and are occurring independently of climate change. When adding the climate change factor, losses can only be expected to increase further. Sound environmental planning is the better economic approach to both sources of loss, while providing better protection of human lives, particularly in typhoon and flooding prone regions. The transitional ambiance in Vietnamese rural areas only exacerbates the problem. For example, the most successful of agriculture and aquaculture farmers, forestry entrepreneurs, gold miners, industrialists and officials may hastily generate enough wealth to leave their local areas, and many do in fact leave for a better life in the city or abroad. But a consequence is that they take little interest in the environment they leave behind for their less fortunate community members. Not pushing economic opportunities to the limit may be a better way to secure income opportunities for the remaining inhabitants and may provide a sustainable environment for future generations.

Third, there is an ethical dimension. Placing climate change in a longer time frame, involving potentially far greater impacts than presently felt, inevitably brings 'global justice' to the fore and calls for assistance to developing countries. Unfortunately, this is not enough to safeguard rural livelihoods. Since climate disruptions manifest themselves through both weather events and the state of the natural environment in which all take a share in a given location, the pattern of distribution of vital resources and entitlements should be the accompanying concern. It is consistent with both the social vulnerability and entitlements approaches to extend the climate change adaptation debate with aspects of social and environmental justice. Equal access to land, water, energy and other resources, equal access to protection from natural hazards and severe weather, equal treatment in government policy and planning, and equal access to participation in policy making and regulation, are as essential for local communities, as the global politics of responsibility and justice. Irrespective of the political context, international mechanism like the 7th Millennium Development Goal of environmental sustainability, and the international environmental and biodiversity protection regimes, should be natural ingredients in any climate change response.

An overall framework for addressing climate change impacts and adaptations would therefore be three-dimensional, consisting of strengthening local community and household robustness, resilience and adaptation (a social dimension); securing overall social, economic and political development (a state dimension); and protecting environment and resources, seeing climate and environment as a unity (an environmental dimension).

7 Methodologies and Common Constraints

This study uses a range of independent and mixed methods, ranging from climate change scenario construction and quantitative and statistical analyses to qualitative interviewing, policy analyses and fieldwork. The separate chapters will outline their respective methodologies.

In Quang Nam province we started out with a comprehensive household survey, including 166 households in 5 districts across the province, and covering a broad range of issues related to socio-economic conditions, stress factors, land uses and changes, climate responses, past disaster experiences, local government responses and health impacts. Extensive data-checking and follow-up interviewing in all communities created a large knowledge base as well as a valuable network for continued research, both among local people, officials and institutions.

Concerning current land uses and land use changes in Central Vietnam under the influence of external factors like policy making, the market, and climate change, the project in Quang Nam built up a large amount of data spanning both agriculture, aquaculture and forestry. Separating the various impacts on current land use changes and new production strategies is inherently difficult, due to methods to distinguish climate change impacts needing further development. However, ongoing analyses with inputs from hydrological studies, biomass and crop analysis, economics, disaster management studies and sociology, including migration and ongoing changes in ownership structures, all address the central issues of vulnerability, resilience capacity and adaptation.

Working with government is a sine-qua-non in Vietnam and access was not always as easy as we wished. It required lengthy procedures through our counterpart in Hanoi and subsequently through provincial and local governments, where there was always a possibility of each level having their own reservations. Certain districts in Quang Nam province are not easily accessible for research, despite the entire province in principle being open to tourism. Particularly, highland districts with large minority population segments proved difficult to access. For this reason the present study has a bias towards lowland and mid-elevation areas and ethnic Vietnamese communities (except "Interacting Cultural and Environmental Change: The Co (Cua) Minority of Central Vietnam"). Similarly, the countless localities for hydropower construction were in principle out of bounds. Considering the current enormous scale of funding, authorities may not be impartial to the exposure of external climate threats to anyone associated with foreign donors. Corruption is generally acknowledged to remain at a high level, such as in land management and emergency aid (e.g. WB 2011) despite constant government programs to combat it. As a consequence, a range of economic activities and land and resource use practices escape formal statistics, while infusing any data material with great uncertainty.

The current state of statistical data accumulation in Vietnam is a separate issue, as a great local diversity of practices and capabilities contribute to statistical uncertainties. A systemic challenge is the cumulative method of reporting, by means of which data travel from Village to Commune to District to Province authorities (through four independent levels), each level working up the statistics and passing it on to the next level, at the same time having their own separate policy implementation and quota duties. When data reaches the statistical bureau there is a probability that it is already more cooked than raw, so to speak. According to interviews with statistical staff, from a professional point of view a direct mode of reporting from lowest to highest level would ensure a higher quality of data. For instance, data on forestry may stem from various departments and administrative levels using their own methods, such as discussed in "The Push for Plantations: Drivers, Rationales and Social Vulnerability in Quang Nam Province, Vietnam".

Studies of social climate change impacts would automatically seek to establish baselines, or baseline scenarios, against which impacts are measured, with or without adaptation. However, for the reasons above, even when consisting of dynamic projections of land use, production, population growth, urbanization, and so forth, they tend not to capture the true scale of change as experienced on the ground. The problem is how to proceed under these circumstances? Working up statistics, conducting research in local communities, taking samples, etc. is the conventional approach, which may, over time, produce the desired results; several contributions to this book follow this track. Another approach, perhaps more true to experience, would be to reject baselines and apply instead a series of momentary analyses, focusing on conditions, practices, allocations, needs and inequalities at any given point in time. A truly dynamic approach will necessitate repeated field studies and analyses, however, possibly involving local participation in monitoring change. While the first approach may favor larger scale analyses and state-driven interventions, the latter may favor the smaller scale issues, taking into account the totality of impacts from climate and environmental change as well as from general development.

The multiple, interacting processes of change that impact any household in Central Vietnam are at the same time manifestations of those global processes that tie the region to the outside world. The people of Central Vietnam stand on the edge of a new frontier—the frontier of unfulfilled dreams and great opportunities, but at the same time of many perils and unsolved problems.

References

ADB (2009) (Asian Development Bank) The economics of climate change in Southeast Asia: a regional review. ADB, Mandaluyong City

Adger WN (1998) Indicators of social and economic vulnerability to climate change in Vietnam. CSERGE Working Paper GEC 98-02

Adger WN (2006) Vulnerability. Tyndall Centre for Climate Research, Norwich

Adger WN, Kelly PM (1999) Social vulnerability to climate change and the architecture of entitlements. Mitig Adapt Strat Glob Change 4:253–266

Bankoff G (2003) Constructing vulnerability: the historical, natural and social generation of flooding in Metropolitan Manila. Disasters 27(3):224–238

Berkes F, Folke C (eds) (1998) Linking social and ecological systems. Management practices and social mechanisms for building resilience. Cambridge University Press, Cambridge

Berkes F et al (2003) Navigating social-ecological systems: building resilience for complexity and change. Cambridge University Press, Cambridge

Birkmann J (2011) First- and second-order adaptation to natural hazards and extreme events in the context of climate change. Nat Hazards 58:811–840

Boateng I (2012) GIS assessments of coastal vulnerability to climate change and coastal adaptation planning in Vietnam. J Coast Conserv 16:25–36

Brooks N (2003) Vulnerability, risk and adaptation: a conceptual framework. Tyndall Centre for Climate Change Research, Working Paper 38

Bruun O (2012) Sending the right bill to the right people: climate change, environmental degradation and social vulnerabilities in Central Vietnam. Weather, Clim Soc 4(4):250–262

Chaudhuri S et al (2002) Assessing household vulnerability to poverty from cross-sectional data: a methodology and estimates from Indonesia. Department of Economics, Columbia University, Discussion Paper 0102-52

Chaudhry P, Ruysschaert G (2007) Climate change and Human Development. UNDP, Human Development Report Office Occasional Paper, 2007/46

Cutter S et al (2003) Social vulnerability to environmental hazards. Soc Sci Quart 84:242–261

Dao N (2010) Dam development in Vietnam: the evolution of dam-induced resettlement policy. Water Altern 3:324–340

Eriksen S et al (2011) When not every response to climate change is a good one: Identifying principles for sustainable adaptation. Clim Dev 3:7–20 Review article

Fabres B (2011) Think global, act global in the Mekong Delta? Environmental change, civil society and NGOs. In: Stewart MA, Coclanis PA (eds) Environmental change and agricultural sustainability in the Mekong Delta. Springer, Dordrecht, pp 7–34

Folke C (2006) Resilience: the emergence of a perspective for social-ecological systems analysis. Glob Environ Change 16(3):253–267

Fortier F (2010) Taking a climate chance; a procedural critique of Vietnam's climate change strategy. Asia Pac Viewpoint 51(3):229–247
Füssel HM, Klein RT (2006) Climate change vulnerability assessments: an evolution of conceptual thinking. Clim Change 75:301–329
Gaillard JC et al (2007) "Natural" disaster? A retrospect into the causes of the late-2004 typhoon disaster in Eastern Luzon, Philippines'. Environ Hazards 7(4):257–270
Giddens A (2009) The politics of climate change. Polity Press, Cambridge
Hewitt K (ed) (1983) Interpretations of calamity: from the viewpoint of Human ecology. Allen and Unwin, Boston
Holling CS (1973) Resilience and stability of ecological systems. Annu Rev Ecol Syst 4:1–23
LAV (2009a) (Look at Vietnam) Numerous hydroelectric plants built in the central highlands region have made floods in the storm-hit area more serious, experts say. www.lookatvietnam.com. 12 December
LAV (2009b) (Look at Vietnam) Deputy Minister blames reservoirs for worsening floods. www.lookatvietnam.com. 10 October
Marino E, Ribot J (2012) Special issue introduction: adding insult to injury: climate change and 5th inequities of climate intervention. Glob Environ Change 22:323–328
McElwee P (2010) The social dimensions of adaptation to climate change in Vietnam. The World Bank, Discussion Paper 17, December
Mitchell M (2009) Complexity theory: a guided tour. Oxford University Press, Oxford
MONRE (Ministry of Natural Resources and Environment) (2007) National target program to respond to climate change. MONRE, Hanoi
MONRE (Ministry of Natural Resources and Environment)— (2009) Climate change: sea level rise scenarios for Vietnam. MONRE, Hanoi
Moser C (1998) The asset vulnerability framework: reassessing urban poverty reduction strategies. World Dev 26(1):1–19
Nelson DR et al (2007) Adaptation to environmental change: contributions of a resilience framework. Ann Rev Environ Res 32:395–419
O'Brien KL, Leichenko RM (2000) Double exposure: assessing the impacts of climate change within the context of economic globalization. Glob Environ Change 10:221–232
O'Brien KL et al (2004) What's in a word? Conflicting interpretations of vulnerability in climate change research. CICERO Working Paper
OECD (2009) OECD policy guidance on integrating climate change adaptation into development co-operation. OECD, Paris
Prigogine I (1997) The end of certainty. The Free Press, New York
Schad I et al (2012) Why do people not learn from flood disasters? Evidence from Vietnam's northwestern mountains. Nat Hazards 62:221–241
Schipper L (2007) Climate change adaptation and development: exploring the linkages. Tyndall Centre for Climate Change Research, Working Paper 107
Scott JC (2004) Seeing like a state: how certain schemes to improve the human condition have failed. Yale University Press, New Haven
Scott JC (2009) The art of not being governed: an anarchist history of upland Southeast Asia. Yale University Press, New Haven
Sebesvari Z et al (2011) Climate change adaptation and agrichemicals in the Mekong Delta, Vietnam. In: Stewart MA, Coclanis PA (eds) Environmental change and agricultural sustainability in the Mekong Delta. Springer, Dordrecht, pp 219–240
Sperling F et al (ed) (2002) Poverty and climate change: reducing the vulnerability of the poor through adaptation. African Development Bank
Van Tan P (2010) Study on developing detailed climate change scenarios for mid-central Vietnam. Sub-project 3, project 'Assessing impacts of climate change on physical, environmental conditions and socio-economic development in mid-central Vietnam'. Hanoi, Institute of Geography, Vietnamese-Danish research project
WB (2010) (World Bank) Economics of adaptation to climate change: Vietnam. World Bank

WB (World Bank with Danida and Sida) (2011) Recognizing and reducing corruption risks in land management in Vietnam. National Political Publishing House, Hanoi

Williamson T, Hesseln H, Johnston M (2012) Adaptive capacity deficits and adaptive capacity of economic systems in climate change vulnerability assessment. For Policy Econ 24:48–54

Wisner B et al (2004) At risk: natural Hazards, People's vulnerability and disasters, 2nd edn. Routledge, London

Paradoxes in Adaptation: Economic Growth and Socio-Economic Differentiation. A Case Study of Mid-Central Vietnam

Mogens Buch-Hansen, Nguyen Ngoc Khanh and Nguyen Hong Anh

Abstract This chapter will introduce the geographical conditions and economic characteristics of Central Vietnam. It proceeds to analyse the general economic development in the region, particularly since the introduction of economic reforms in the early 1990s. Both foreign investment and exports have risen dramatically, turning Vietnam into a major exporter of a range of agricultural and industrial products. Further, it shows how some of the paradoxes of development and globalization, such as high growth and simultaneous socio-economic differentiation, are also played out in the provinces of Central Vietnam. In a disaster prone region, the poor households are at risk of losing out when greater weather variability threatens agricultural and forestry production and increases overall economic losses. The rapid expansion of the hydropower generating capacity has stimulated economic development and thereby potentially enhanced social resilience, but at the same time has increased the ecological vulnerability and set in motion a range of processes not under control.

Keywords Economic development • Social differentiation • Adaptation capability • Social resilience and ecological vulnerability

1 Introduction

For nearly two decades now, international debates at the COP meetings have centred on the costs of mitigating climate change among the big greenhouse gas (GHG) emitters in the 'old' industrialised countries. They have also addressed the problem of escalating emissions among the emerging economies, and the costs of

M. Buch-Hansen (✉)
Department of Society and Globalisation, Roskilde University, Roskilde, Denmark
e-mail: mobh@ruc.dk

N. N. Khanh · N. H. Anh
Institute of Sustainable Development, Hanoi, Vietnam

O. Bruun and T. Casse (eds.), *On the Frontiers of Climate and Environmental Change*, Environmental Science and Engineering, DOI: 10.1007/978-3-642-35804-3_2,

adapting to the inevitable changes resulting from climate change.[1] Climate change has become part of wider debates on climate (in)justice and rights to development (Huq et al. 2006) as a result of confirmation by the International Panel for Climate Change (IPCC) that climate changes are induced by humans (IPCC 2007), and the 2006 Stern Report's estimation of the costs of climate changes (Stern 2006). Another important issue debated at the COP meetings concerns how to share the costs of adaptation to ongoing climate changes, especially those incurred by poorer developing countries (World Bank 2010b). Vietnam is highly vulnerable to predicted increases in climate variability and extremes, making this an issue of major concern to the political leaders at the helm of the country's rapid economic growth. Vietnam's long coastline bordering the Chinese Sea, the vast low-lying river deltas of the Mekong and the Red River, and the steep slopes between the high mountains bordering Lao PDR and the coastline in Mid-Central Vietnam make Vietnam very prone to rising sea levels, heavy storms and the extreme precipitation resulting from frequent tropical typhoons. Even leaving aside the predicted increase in climate variability and extremes, current extreme weather conditions (including frequent typhoons with excessive rains and heavy storms, an unpredictable farming calendar, salt water intrusion, etc.) are seriously undermining economic development and accentuating social differentiation.

However, in order to fully understand how climate-related natural disasters affect environments and economic development, it is essential to investigate the relationship between climate variability and extremes, and man-made environmental changes (Bruun 2012). The World Bank's Discussion Paper on the Social Dimensions of Adaptation to Climate Change in Vietnam states that between 1953 and 2009 climate related natural disasters resulted in nearly 25,000 deaths, with a further 77 million people affected. Damages due to natural disasters during the same period have been estimated at more than 7 billion USD (World Bank 2010a). This chapter will analyse economic development and socio-economic differentiation in Mid-Central Vietnam, and specifically Quang Nam Province, which is one of the provinces that is highly vulnerable to predicted increases in climate variability and extremes which are likely to exacerbate existing problems.

1.1 The Transition After Reunification

During the decade following reunification in 1975, the Vietnamese economy was busy recovering from many years of war, a process which both involved rebuilding the country's physical infrastructure and restoring its social, economic and institutional infrastructure. The second Five Year Plan (1976–1980) called for a major emphasis on heavy industry and rapid agricultural growth.

[1] COP stands for Convention of Parties which is the annual meeting held among the parties of the UN Framework Convention for Climate Change (UNFCCC).

> The optimism and impatience of Vietnam's leaders were evident in the Second Five Year Plan. The plan set extraordinarily high goals for the average annual growth rates for industry (16–18 %), agriculture (8–10 %), and national income (13–14 %). It also gave priority to reconstruction and new construction while attempting to develop agricultural resources, to integrate the North and the South (The Library of Congress Country Studies 1987).

During this period, industrial production grew by a mere 0.6 % and agriculture by 1.9 %, and as such the Plan must be considered a failure. This failure was due to a number of adverse factors such as problems in reunifying the economy, unfavourable weather conditions, high military expenditure, managerial inefficiency, etc. However, the poor performance of the productive sectors seems to have been compensated for by deforestation and logging activities, where peak output was attained during the Second Five Year Plan (Beresford and Fraser 1992, p. 7). Following the 6th National Party Congress in 1986, Vietnam embarked on its renewal process, *Doi Moi*, which marked the beginning of a period of unprecedented economic growth and social dynamism throughout the entire Vietnamese economy. Vietnam thus became an example of a thriving socialist oriented market economy whose progress was only briefly slowed by the Asian financial crisis of 1997 and the global economic crisis that occurred just over a decade later in 2008.

This chapter describes some of the main achievements of the country as a whole, and also focuses specifically on the South Central Coastal Region and Quang Nam Province. The aim is to show how the economic growth and socio-economic differentiation that have occurred since the introduction of *Doi Moi* have created paradoxes in the capacity of particular communities and socio-economic groups to adapt to climate variability and extremes in the region. The South Central Coastal Region of Vietnam is highly vulnerable to predicted increases in climate variability and extremes (World Bank 2010b; Kelly et al. 2001). Yet existing climate conditions are already extreme, such as frequent typhoons, salt water intrusion from rising sea levels, heat waves and periodic droughts, calling for dynamic adaptation measures to protect the economy and livelihoods of the local communities.

Vietnam's rapid economic growth since the mid-1980s has greatly enhanced the country's overall resilience to extreme weather conditions, but it has also produced a number of paradoxes, notably differences in various groups' capacity to adapt to increased climate variability and extremes. We address this issue below, drawing on the concepts of vulnerability, resilience and adaptation. We use resilience and vulnerability as antonyms, where resilience refers to the number of external disturbances a system can withstand before it is drastically or even irreversibly changed. Adaptation is, in turn, understood here as activities aimed at decreasing vulnerability to such external disturbances (Adger et al. 2001; Ensor 2011).

2 Doi Moi and Rapid Economic Growth

A number of laws were implemented in the years following the 6th National Party Congress with a view to facilitating the transition to a market economy and attracting and securing foreign investments. These laws include the Foreign Investment

Law of 1987, the Private Enterprise Law and Corporate Law introduced in 1991, followed by the amendment of the Constitution in 1992. These laws all affirmed the existence and development of a multi-sector, market economy, including foreign investments. They were followed by the Land Law of 1993, the Environment Law (1993), the Bankruptcy Law (2004) and the Intellectual Property Law (2005), as well as hundreds of ordinances and decrees that were enacted to guide the implementation of the socialist oriented market economy.

The results of these reforms in terms of economic growth are impressive, as shown in Fig. 1. From a relatively slow start, growth in GDP of 2.8 % p.a. in 1986 rose to 8.2 % p.a. for the period 1991–1995 and then dropped to 7.5 % p.a. from 1996 to 2000 due to the Asian financial crisis. Thereafter, it slowly picked up again, reaching 8.5 % in 2005. Vietnam's GDP decreased following the 2008 global financial crisis, but picked up much faster than most other economies, peaking at 6.8 % in 2010 and then dropping again to 5.9 % in 2011. These high levels of economic growth were triggered both by the transition of the economy from agriculture to industry, trade and services, and the privatisation of the economy, including the influx of foreign investments. Table 1 shows the transition in the structure of the economy during the 20 years following the introduction of *Doi Moi* in 1986, with decreasing contributions to GDP from agriculture, forestry and fishing and increasing contributions from industry and construction. The service sector's contribution to GDP has remained relatively constant, at 38.6 % in 1990, 38.1 % in 2005 and 38.3 % in 2010. However, the high quality services within finance, insurance, IT and tourism increased rapidly. The number of foreign tourists visiting Vietnam increased by 63 % between 1999 and 2004 from 1.7 million to 2.9 million; and then again by 45 % to 4.2 million in 2008. Due to the global financial crisis, the number of tourists fell to 3.8 million in 2009 but the total number of foreign visitors to Vietnam in 2010 still exceeded 5 million.

Economic growth since the introduction of *Doi Moi* has, to a large extent, been spearheaded by the economic privatisation process, not least the growing foreign investments. From 1991 to 2003, the private sector's share of GDP rose from 3.1 to 4.1 % while the GDP share of other non-state sectors increased from 4.4 to 4.5 %. Meanwhile, the foreign investment sector grew from 6.4 to 14 %. With the promulgation of the Enterprise Law in 2000, the number of registered private enterprises more than doubled, rocketing from 73,000 to 150,000 in 2004. The inflow of private investments ranged between 1.3 billion and 1.8 billion USD per annum between 2002 and 2006, while in 2007 they soared to 6.6 billion USD and then to 9.3 billion USD in 2007. Throughout the same period, exports grew steadily at around 20 % per year, reaching 26 billion USD in 2004, 32.23 billion USD in 2005 and 56.6 billion USD in 2009. Of these totals, 47 % resulted from FDI (General Statistical Office of Vietnam). The global financial crisis caused a drastic drop in inward FDI in 2008 and 2009 but had already picked up by 2010 and was expected to increase further by 2011[2] as Vietnam is considered an attractive

[2] Figures for 2011 are not yet available.

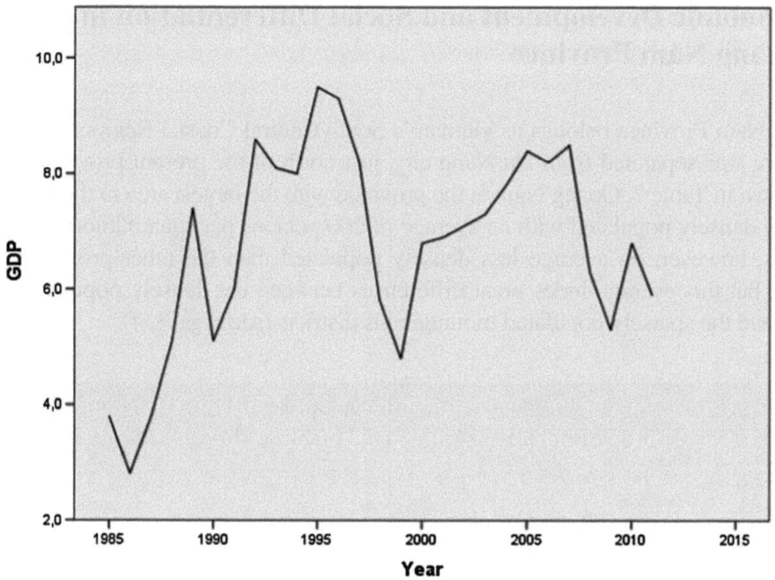

Fig. 1 GDP growth, 1985–2011

Table 1 Structure of sectors' contributions to Vietnam's GDP

Sector	1990	1995	2005	2010
Agriculture, forestry and fishing	38.7 %	27.2 %	20.9 %	20.6 %
Industry and construction	22.7 %	28.8 %	41.0 %	41.1 %
Services	38.6 %	–	38.1 %	38.3 %
Total	100		100	100

Source Vietnam statistical yearbooks

location for FDI in Southeast Asia. Like most of the Asian emerging economies, Vietnam displays very high rates of investment, amounting to 33.9 % of GDP in 2009 (General Statistical Office of Vietnam). Yet the picture is not all rosy—in the conclusion We touch on some recent problems afflicting the Vietnamese economy.

The agricultural sector has by no means been left behind by the rapid growth of the industrial sector, especially not compared to the pre *Doi Moi* era when Vietnam experienced long periods of hunger and malnutrition. The promotion of agriculture and aquaculture has made Vietnam self-sufficient in basic foodstuffs and a prime exporter of agricultural and aquaculture commodities, making it the world's biggest exporter of cashew nuts in 2010, the second largest producer of rice and coffee, fourth in rubber exports, and fifth in fishery and aquaculture, with cultivated shrimps the fastest growing export commodity.

A strong indicator of the strength of *Doi Moi* is the fact that GDP per capita more than quadrupled in only twenty years in current prices, from 200 USD in 1990, to 640 USD in 2005, to 1.100 USD in 2010.

3 Economic Development and Social Differentiation in Quang Nam Province

Quang Nam Province belongs to Vietnam's South Central Coastal Region. In 1995, the province was separated from Da Nang city, just north of the present provincial border. As shown in Table 2, Quang Nam is the province with the largest area in the region, and is fairly densely populated with an average of 200 persons per square kilometre. Quang Nam is, however, on average less densely populated than the other provinces in the region, but this average hides great differences between the densely populated coastal plains and the sparsely populated mountainous districts (see Figs. 3, 4).

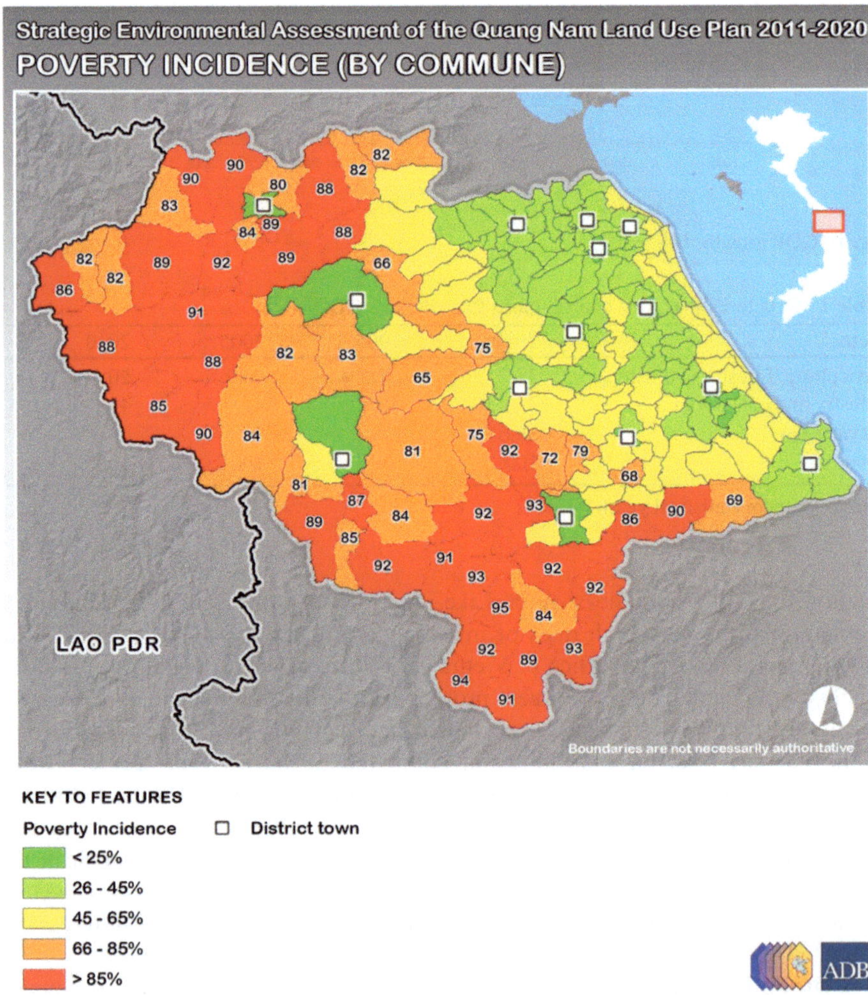

Fig. 2 Poverty incidence in Quang Nam Province (Quang Nam Provincial People's Committee 2010)

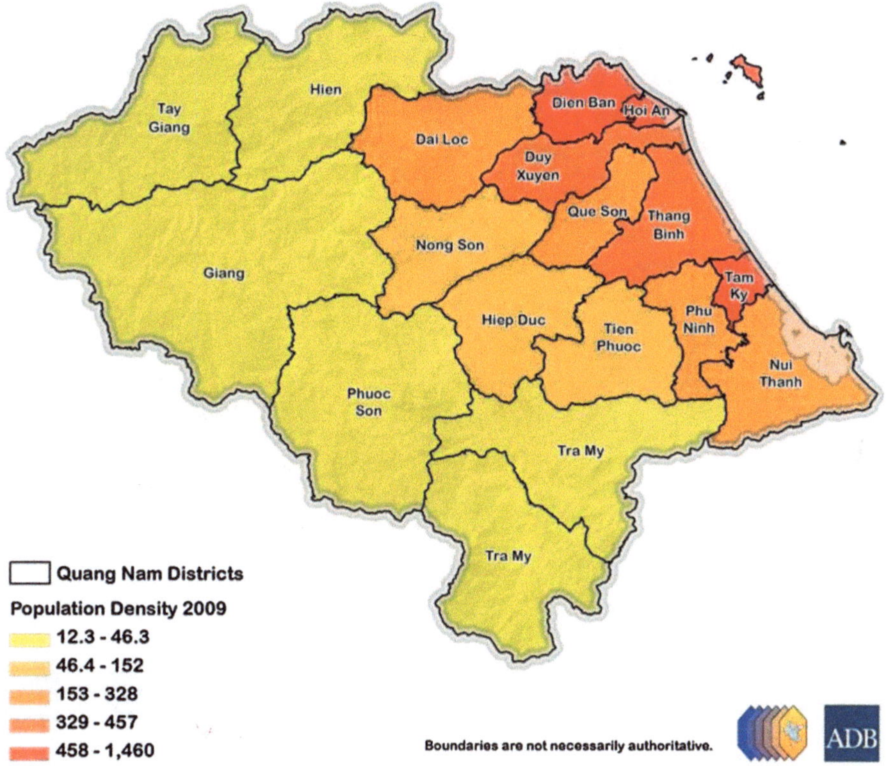

Fig. 3 Population density in Quang Nam province 2009 (Quang Nam Provincial People's Committee 2010)

Table 2 Basic characteristics of the south central coastal region in 2007

No.	Province/ municipality	Area (km²)	Population (2007)	Population density (persons/ km²)	GDP per capita (million VND, 2007)
1	Đà Nẵng	1,256	805,400	641	18.98
2	Quảng Nam	10,438	1,484,300	142	8.76
3	Quảng Ngãi	5,153	1,288,900	250	7.82
4	Bình Định	6,040	1,578,900	261	9.57
5	Phú Yên	5,061	880,700	174	8.43
6	Khánh Hòa	5,218	1,147,000	220	16.10
7	Ninh Thuận	3,363	574,800	171	6.66
8	Bình Thuận	7,837	1,170,700	149	11.00
9	Total	44,367	8,930,700	201	10.76

Source General statistical office of Vietnam

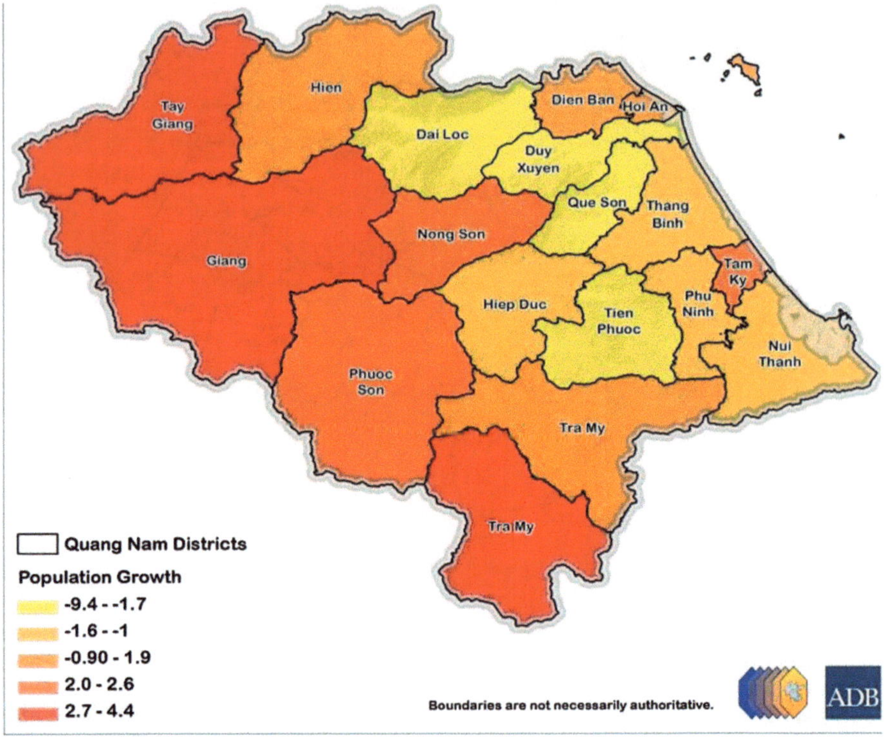

Fig. 4 Average population growth rates 2005–2010, Quang Nam province (Quang Nam Provincial People's Committee 2010)

In 2007, the South Central Coastal Region accounted for 9.5 % of the Vietnamese population and produced 9.7 % of Vietnam's GDP based on agriculture, forestry and fishing. Although considered a relatively highly industrialised region, surpassed only by the industrial hubs of Hanoi and Ho Chi Minh City, the region only accounted for 7.54 % of Vietnam's industrial GDP in 2007 (Vietnam General Statistics Office 2009). Formerly, the region was largely dependent on the direct utilisation of natural resources in agriculture and fishing, which accounted for 47.5 % of the regional GDP in 1990, but by 2002 this dependence on natural resources had been greatly reduced, to 38.7 % of regional GDP. This downward trend continued, falling to a mere 9.5 % by 2010. During the same period, the industry and construction sectors increased their share of GDP from 22.7 % in 1990 to 23.9 % in 2002, further rising to 46.1 % in 2010. Meanwhile, during the same period the service sector grew from 29.8 to 37.4 % and again to 44.4 % of regional GDP in 2010. The sharp increase in contribution to regional GDP by industry, construction and services was due in no small measure to the rapid industrialisation and modernisation of Da Nang, Vietnam's third largest city, which attracted large numbers of tourists to its long white beaches and the nearby charming trading post of Hoi An where centuries earlier Chinese and Portuguese traders had created an East–West cultural melting pot.

The province of Quang Nam covers the largest area of the eight provinces in the South Central Coastal Region, but has a relatively low GDP per capita compared to the regional average, as shown in Table 2. This is largely due to the great geographical diversity that characterises the province, which ranges from the mountainous districts bordering Lao PDR, to the Midlands and the coastal plains. While the coastal plains are just as well off as the coastal plains in the other provinces, the mountainous districts, which are largely inhabited by ethnic minorities, have much larger populations living below the poverty line (see Fig. 2). Being sparsely populated, they are also the districts with the highest rates of population growth, as shown in Figs. 3 and 4. So, while the average provincial natural growth rate fell by more than one half from 1.9 % in 1996 to 0.9 % in 2010, the natural growth rate in the mountainous districts with a higher percentage of ethnic minorities remained high. With growing pressure on land on the coastal plains there has also been an influx of ethnic Kinh people (Vietnam's main ethnic group) who have established their own villages in the mountainous districts.

Tables 1 and 3 show that the transition process began later in Quang Nam Province than in other regions in Vietnam. While in Quang Nam in 2005 31 % of GDP (down from 51 % in 1996) was still generated by agriculture, forestry and fishing, compared to an average 20.9 % for Vietnam in general; and only 34 % of GDP was produced by industry and construction (though this had doubled since 1996!) compared to 41 % on average in Vietnam, it is remarkable that by 2010 the figures for Quang Nam Province had almost caught up with the Vietnamese average. Since the onset of the global financial recession, the province has experienced a very rapid transition from being very dependent on the direct utilisation of natural resources, to becoming more dependent on manufacturing industry, construction and services. As evident from Table 4, although approximately 50,000 jobs

Table 3 Structure of sectors' contributions to Quang Nam's GDP

Sector	1996 (%)	2005 (%)	2010 (%)
Agriculture, forestry and fishing	51	31	21
Industry and construction	17	34	40
Services	32	35	39
Total	100	100	100

Source Quang Nam statistical yearbooks

Table 4 Employees by economic sector in Quang Nam province, 1976–2010

Sector	1976	1985	Growth in % 1976–1985	1990	2000	2005	2010	Growth in % 1985–2010
Agriculture, forestry and fishing	292.120	398.925	36.6	456.880	558.257	531.889	481.112	20.6
Industry and construction	30.793	34.212	11.1	37.592	56.903	88.398	156.892	358.6
Services	3.410	36.456	969.1	37.334	73.135	126.188	174.192	377.8
Total	354.323	469.593	32.5	531.696	688.295	746.475	812.196	73.0

Source Quang Nam statistical yearbooks

were lost within agriculture, forestry and fishing between 2005 and 2010, 116,500 new jobs were created within industry, construction and services during the same period.

While the contribution of agriculture, forestry and fishing to provincial GDP dropped from 51 % in 1996 to only 21 % in 2010, the number of employees in the sector dropped at a much slower pace, which has resulted both in social and income differentiation. This differentiation is seriously impairing different population groups' ability to adapt to the extreme weather events already occurring and will increase the challenges for those groups, considering the predicted increases in climate variability and extremes.

As it is the industrial sector that is the driving force behind the current transition, both in Vietnamese society generally and in the economy of Quang Nam Province, it is interesting to note the change in ownership of gross industrial output during the rapid transition period, as shown in Table 5. The state sector is retreating rapidly, while private investments, and not least foreign investments, are growing swiftly. The global financial crisis is, however, clearly reflected in the slowdown of output from foreign investments in 2009, which had nonetheless already regained momentum by 2010. The South Central Coastal Region, including Quang Nam Province, is playing a major role in Vietnam's drive to integrate itself into the global economy and is, like other regions in Southeast Asia, among the emerging economies that are proving a serious competitor to the old industrial centres of Europe, America and Japan.

Foreign investments are, to a large extent, targeting the new industrial zones in Dian Nam in Dien Ban District between Da Nang and Hoi An, and the very new industrial processing zone of Chu Lai developing in the coastal plain region

Table 5 Industrial gross output (current prices) by ownership in percentage

Quang Nam province	2006	2007	2008	2009	2010 (est.)
State	20.44	13.18	11.21	11.07	9.92
Central	11.42	8.64	7.35	8.46	7.60
Local	9.02	4.54	3.86	2.61	2.32
Non-state	68.44	70.99	69.95	76.17	74.77
Collective	4.07	2.64	1.78	1.01	0.93
Private	49.20	55.96	56.36	64.95	63.33
Households	15.17	12.38	11.81	10.21	10.51
Foreign investments	11.12	15.83	18.84	12.76	15.31
Total	100	100	100	100	100
Dien Ban District[a]	41.6 %	38.6 %	35.2 %	29.1 %	28.0 %
Nui Thanh District	9.0 %	9.0 %	10.9 %	14.0 %	16.4 %
Number of foreign visitors to the Province	653.264	1.102.193	1.143.833	960.830	1.062.982

Source Quang Nam statistical yearbook 2010

[a] The figures for the two districts of Dien Ban and Nui Thanh show the districts' share of Quang Nam Province's Non-State industrial gross output at constant prices

Table 6 Number of tourists to Quang Nam province 1997–2010

Number of tourists/year	1997	2000	2005	2006	2008	2010 (prel.)
International	74,855	218,871	672,554	653,264	1,143,833	1,062,982
Domestic	19,508	163,834	533,372	631,471	620, 383	701,368
Total	94,363	382,705	1,205,926	1,284,735	1,764,216	1,764,350

Source Quang Nam 10 years and statistical yearbook 2010

of Nui Thanh District. In particular, Chu Lai Industrial Zone is attracting a large number of younger men from neighbouring districts and, according to interviews with district and commune officers from the Midlands and coastal plains, off-farm incomes from industrial work account for approximately 20 % of household incomes in the Midlands and coastal plains of the province.

Furthermore, the number of foreign visitors is increasing rapidly (see Table 6). Tourism is the fastest growing sector in the world, and Vietnam in general and the province of Quang Nam in particular are working hard to benefit from that growth, not least by constructing huge tourist resorts at the famous China Beach between Da Nang and Hoi An. Also remarkable is the rapid growth of domestic tourism. This was almost non-existent up to 1997, but in 2007 over 700,000 domestic tourists visited the province. The number of foreign visitors also multiplied by a factor of 14 between 1997 and 2010, and has seemingly only been mildly affected by the 2007 global financial crisis. By 2008, more than 1.1 million foreign tourists visited Quang Nam and it is estimated that over 1 million foreign tourists visited Quang Nam in 2010, accounting for approximately 20 % of the total number visiting Vietnam.

In 2011, the Quang Nam People's Committee in collaboration with UNESCO developed a new integrated culture and tourism strategy for sustainable development that aimed to change the quantitative growth approach to tourism to a quality-based growth strategy. The strategy aims to ensure the sustainability of Quang Nam's culture-tourism sector, including protecting cultural and natural resources, increasing local communities' benefits from tourism, and diversifying provincial tourism products. Special management plans have been drawn up for the World Heritage Sites of Hoi An, My Son and the Cham Island Biosphere Reserve (Duong 2011).[3]

3.1 *Hydropower Production in Quang Nam*

As a rapidly growing emerging economy Vietnam needs power, and hydropower already accounts for approximately one third of Vietnam's power production. It is estimated that demand is increasing by 15 % per year and insufficient

[3] For a critical review of the 'Western' concept of Heritages Sites, in view of the rapidly increasing number of local tourists, see (Hitchcock et al. Hitchcock et al. 2009).

power is by many observers seen as a serious impediment to further economic growth (International Centre for Environmental Management 2006). The Central Highlands have great potential for hydropower generation, and the mountainous districts of Quang Nam Province already host a number of hydropower projects. A further 60 new projects have just been completed, are under construction or are scheduled in the Quang Nam Power Development Plan 2006–2010. This will add about 1,600 MW to the projects already in operation (DOIT: Quang Nam Power Development Plan 2006–2010 with a Vision to 2015). Hydropower projects are, however, a controversial topic in Vietnam, due to contradictions between the dire need for power to fuel the growing economy, and the serious social and environmental impacts of most of these projects. Due to the uncertainty created by contradictory interests, the planning of hydropower projects is somewhat confusing as projects that have already been approved are being cancelled while new ones are constantly being added, requiring new environmental impact assessments. In 2006, a Strategic Environmental Assessment identified 82 environmental, social and economic issues to be dealt with in the Quang Nam Power Development Plan (International Centre for Environmental Management 2006). According to the VietNamNet website, in 2012 Quang Nam Province had 43 hydro-power plants, including ten large scale ones. The three largest works are Song Tranh 2, with a reservoir of 740 million cubic meters of water, A Vuong with 343 million cubic meters and DaMil with 310 million cubic meters (VietNamNet 2012b).

The environmental problems caused by hydropower projects are manifold, especially if the projects require reservoirs to store water for a constant annual production of power. Local people (often ethnic minorities) have to be resettled, and subsequently demand reasonable compensation for lost livelihoods—though they do not always receive this. In addition, endangered species of plants and animals are, in many cases, threatened by the inundation of pristine forests that takes place when creating the water reservoirs. The remaining forest cover around the watershed has to be maintained to secure the infiltration of rain water and to avoid soil erosion that will result in the siltation of the reservoirs. This is often in direct conflict with the local residents who have been evicted from their land to make room for the reservoirs. In order to secure their livelihoods they are forced to clear forest, often in the nearby watershed of their inundated homelands. On top of this, illegal lumberjacks can now enter the hitherto impenetrable forests by use of the waterways provided by the reservoirs. Thus, rare and precious forests that were previously considered inaccessible are now been logged for illegal commercial purposes (VietNamNet 2012a).

In extreme weather situations with excessive rain, proper management of the reservoirs is essential to avoid the water from the reservoirs contributing to the potential flooding downstream when the gates are opened. Similarly, during periods of drought, it is essential that the hydropower projects do not store the water for power generation instead of releasing it for irrigation. Deforestation resulting from the building of the water reservoirs and/or the resettlement of evicted residents, together with reservoir management during excessive rain and drought, are examples of man-made environmental changes that aggravate the environmental and socio-economic impacts of extreme weather situations.

The Vu Gia Thu Bon river system has great potential for hydropower production, and existing projects [A Vuong, Song Tranh, Dak M, Song Con and others (see map, Fig. 5)] already provide more than 1,000 MW.

According to the SEA of the proposed and planned hydropower projects, there are a large number of unsolved environmental, social and economic issues that occasionally lead to some projects being scrapped in favour of new potential projects. The demand for power to fuel the rapidly growing economy is huge, leading to intense political pressure to implement hydropower projects despite the above-mentioned environmental, social and economic problems. The political power of the stakeholders involved is highly unequal. The ethnic minorities and small-scale farmers being resettled, those whose fields are inundated during excessive rains due to poor management of the water reservoirs, and those who lack water for irrigation during periods of drought due to the prioritisation of hydropower production, are politically weak compared to the ruling elites and private investors who need the electric power. When confronted with the question of how one should prioritise between using the water in the reservoirs for irrigation or for hydropower production during periods of drought, high ranking provincial officers in Quang Nam Province unanimously confirmed that hydropower production will be given first priority, although they could definitely see the problems caused by it (Personal communication in October–November, 2011).

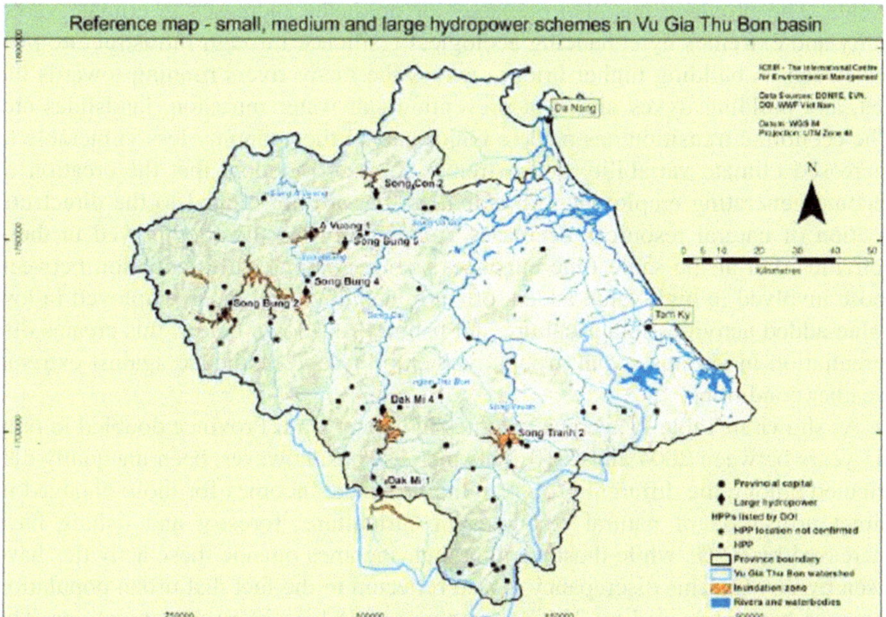

Fig. 5 Map of proposed and existing hydropower schemes in Vu Gia Thu Bon Basin, covering most of Quang Nam province

4 Economic Development, Socio-Economic Differentiation and Its Impact on Adaptation Capabilities

National and local communities' vulnerability and resilience towards extreme weather situations has been heavily influenced by the *Doi Moi* economic renewal process, particularly the transition from a resource dependent society to an emerging industrial society with vigorous economic growth and increasing social differentiation. Awareness of these issues should inform any planning initiatives designed to ease adaption to this new situation. In that connection, it is important to distinguish between ecological and social vulnerability and resilience, as well as to highlight their very close relationship (Adger 2000). Whereas resilience, as described above, refers to the amount of external disturbances a system can withstand before it is drastically and even irreversibly changed, adaptation capabilities refer to how a society or smaller community can enhance its ecological resilience (e.g. through changes in land use towards more weather resilient crops or activities, constructing dykes or river protection walls, etc.) and social resilience (e.g. changing from on-farm incomes to off-farm incomes, or changing to high income generating on-farm activities like acacia plantations or shrimp farms, establishing warning systems and assuring disaster preparedness, etc.) (Ensor 2011).

The rapid economic growth that has taken place in Vietnam since the mid-1980s, and in Quang Nam Province since the turn of the millennium, has enhanced both national and provincial social resilience against extreme weather events. The province has also prepared for predicted increases in climate variability and extremes by enhancing ecological resilience through infrastructure projects such as building higher bridges across the many rivers running towards the sea, and building dykes aimed at preventing salt water intrusion, landslides etc. The economic transition has made a major part of the economy less vulnerable to increased climate variability and extremes. It is self-evident that the creation of income generating employment in activities that are not related to the direct utilisation of natural resources decreases vulnerability for those employed in these activities, but at the same time enhances socio-economic differentiation between those involved in high value-added, off-farm activities and those employed in low value-added activities in agriculture and fishing. As shown below, this creates differentiation in various social groups' vulnerability and resilience against extreme weather conditions.

As shown in Table 7, average incomes in Quang Nam Province doubled in only six years between 2004 and 2010. This increase has, however, been unequally distributed among the different groups in the province: incomes for those engaged in direct utilisation of natural resources in agriculture, forestry and fishing have increased by 62 %, while those earning their incomes outside these activities have risen by 264 %. This discrepancy is also reflected in the fact that urban population incomes have increased by 236 % compared to 184 % for rural populations. The 20 % richest households' incomes grew by 233 %, compared to the 20 % poorest households whose incomes rose by 152 %. Although the increase in incomes is

Table 7 Monthly average income per capita (1,000 VND at current prices) by residence, income source and income quintile

	1999	2002	2004	2006	2008	2010 (estimated)	Percentage change 2004–2010 (%)
Total			328.8	459	693.7	985.8	200
Urban			414	626.4	939.2	1393	236
Rural			312.8	424	641.1	886.9	184
Agriculture, forestry and fishing			103.7	120.9	176.6	168.5	62
Non-agriculture, forestry and fishing			60.5	88.4	124.8	220	264
Quintile 1	78.9	104.9	122.4	166.2	249.3	308.9	152
Quintile 5	512.8	503.4	665.4	938.2	1441	2218.9	233

Source Quang Nam statistical yearbook 2010 and 2004

substantial, even for the poorest households, the two groups' perceptions about the extent to which their livelihoods have improved do not reflect this. According to our household survey, 83 % of the richest households perceived that their livelihoods had improved substantially, compared to only 23 % of the poor households. Conversely, 31 % of the poorest households felt that their living standards had deteriorated substantially, which could be explained by the decline in social support for the socially most vulnerable groups.[4]

Paradoxes in Vietnam's strategy to reduce vulnerability towards predicted adverse climatic changes arise from a number of factors: the extreme weather conditions that already prevail in the South Central Coastal Region and Quang Nam Province; the predicted increase in climate variability and extremes; and the rapid transformation from reliance on agriculture, forestry and fishing to becoming an emerging industrial society with rapid economic growth and socio-economic differentiation. These issues will be analysed in the following section.

5 Adaptation Capabilities: Paradoxes Resulting from Economic Growth and Socio-Economic Differentiation

With the inevitable growing demand for energy, and the favourable conditions (seen from a power production point of view) for hydropower generation in the Central Highlands of Vietnam, the number of hydropower projects will undoubtedly increase in the future, as will their accompanying social and environmental problems. Increased power production will fuel further transition of the growing

[4] Results of a household survey of 166 households in five districts in Quang Nam Province, carried out by the research project in 2009–2010.

economy, thereby enhancing social resilience against adverse weather conditions. The paradox is that by enhancing social resilience, the hydropower projects and the management of the water reservoirs also increase ecological vulnerability by exacerbating flooding during excessive rains and making the amount of water available for irrigation scarce during droughts. At the same time, the construction of water reservoirs for the hydropower projects is threatening endangered species of plants and animals, as described above. The clearing of land to make room for water reservoirs simultaneously creates social problems, especially for vulnerable ethnic minorities in the Highlands who have been evicted from their land.

Another paradox stems from socio-economic differentiation, as richer households invest in multi-storeyed concrete houses (83.3 %) and are thereby safer in terms of human security and health during flooding or heavy storms. This contrasts with the poorer households, none of which were living in multi-storey concrete houses and whose life and health was consequently much more at risk during extreme weather events. But, according to the abovementioned household survey, 38.5 % of poorer households have become more dependent on agriculture, and none have become less dependent on it over the last ten years. Conversely, among richer households, 83.3 % had become less dependent on agriculture, or had observed no change. Even though extreme weather events threaten poorer households' life and safety and destroy their crops, the latter regarded lack of access to land and employment as a greater threat to their livelihoods than climate drivers. Rich households, however, perceived climate change as the biggest threat to their livelihoods even though they were safer during extreme weather conditions and less dependent on agriculture, due to the fact that extreme weather events such as storms and excessive rains had the potential to ruin their substantial investments in high income earning activities, such as acacia plantations in the Midlands and Highlands and shrimp farms in the Lowlands (Buch-Hansen 2013).

6 Conclusions

The rapid economic growth which has taken place within the past 10–15 years following the economic transformation in Vietnam in general since *Doi Moi*, and in Quang Nam Province specifically, has greatly enhanced social resilience against extreme weather conditions. The contribution to the economy by agriculture, forestry and fishery, the sectors most volatile to extreme weather conditions, has dropped to about 20 %, even though more than 50 % of the workforce in Quang Nam Province earn their incomes from these primary production sectors. The increased social differentiation, however, has also led to changes in different social groups' ability to adapt to extreme weather conditions: rich households have built strong, concrete multi-storey houses, making them safer during extreme weather events, while the vast majority of poor, rural households live in poorly constructed mud and brick one-storey houses, vulnerable to storms and floods.

The rapid societal transition currently underway in Vietnam, which has been especially dramatic in Quang Nam Province since the turn of the Millennium, has generated considerable concern about the region's increasing market orientation and market dependency. Legal reforms have also sparked worries about rights and access to land. At the same time, new hydropower projects with huge water reservoirs are springing up all over the province. These, together with widespread deforestation, contribute to major man-made environmental changes which exacerbate the adverse impacts of extreme weather conditions (International Centre for Environmental Management 2006; VietNamNet 2012b). The predicted increase in climate variability and extremes is, however, often perceived as the main future culprit for natural disasters created by extreme weather conditions and hence for slowing down economic development. This development scenario is what creates some of the paradoxes in different groups' ability to adapt to extreme weather situations. On the one hand, hydropower projects greatly contribute to economic growth and thereby enhance social resilience against extreme weather conditions but, on the other hand, they contribute to flooding during heavy precipitation and they reduce the amount of water available for irrigation during droughts (Personal communication with high-ranking officers from Quang Nam Province in October–November, 2011).

The other paradox is that richer households which are more secure in terms of life and health during extreme weather conditions, perceive climate change as a bigger risk to their livelihoods due to the risk of losing major investments in fragile high income generating activities like acacia plantations and shrimp farming. Conversely, poorer households are more concerned with their lack of access to land and employment, which they perceive as a bigger obstacle to improving their livelihoods than climate change drivers, even though they are more vulnerable during extreme weather events.

The above discussion strongly suggests that continued economic growth, and the resulting increase in socio-economic differentiation, will further differentiate various groups' ability to adapt to the predicted increase in climate variability and extremes. Vietnam and Quang Nam Province will be in a better position to invest in infrastructure to enhance ecological resilience, such as better roads and higher bridges that will ensure transport of high value products for the market and assist those who can afford a car. Likewise, building dykes can help protect shrimp farms that are normally constructed close to the coastline and are consequently vulnerable to rising sea levels and heavy storms. Enhancing the capabilities of the poor to adapt to more extreme weather situations thus entails focussing on general development initiatives that will improve their on-farm production and incomes through better access to land and off-farm incomes. This will enhance their social resilience and thereby empower them to adapt to climate changes.

The rapid economic growth that has hitherto enhanced social resilience against extreme climate conditions is, however, being questioned as economic growth is no longer living up to expectations. GDP growth in 2010 and 2011 was only 6.8 and 5.9 % respectively, compared to the Government's prediction of 7–7.5 %. Inflation exceeded 20 % in 2011, giving Vietnam the lead in inflation among Asian countries for the second time in three years. This malaise is commonly attributed to poorly run, corrupt and wasteful state owned enterprises and a general failure to enhance labour productivity, which have prevented Vietnam from moving up the value chain, leaving countries like Cambodia and Bangladesh to undercut Vietnam in cheap manufacturing (The Economist 2012).

As social differentiation is unlikely to diminish soon, and if economic growth slows down seriously, we will see decreasing social resilience against increasing climate variability and extremes, especially among the poorer segments of the population, which will aggravate social differentiation and, in turn, lead to even greater vulnerability to climate extremes.

While Vietnam's contribution to global greenhouse gas emissions is still minor, it must be acknowledged that these are not the sole root causes of natural disasters which are often conveniently blamed on climate changes and international climate negotiations. The environmental and socio-economic impacts of the present extreme weather conditions, as well as the predicted increase in climate variability and extremes, are also very closely related to man-made environmental changes such as deforestation, water reservoir construction, and the destruction of coral reefs and mangrove forests for increased aquaculture production, all matters that are entirely in the hands of the Vietnamese themselves.

References

Adger WN (2000) Social and ecological resilience: are they related? Prog Hum Geogr 24(3):347–364

Adger WN, Kelly PM, Nguyen HN (eds) (2001) Living with environmental change: social vulnerability, adaptation and resilience in Vietnam. Routledge, London

Beresford M, Fraser L (1992) Political economy of the environment in Vietnam. J Contemp Asia 22(1):3–19

Bruun O (2012) Sending the Right Bill to the Right People: Climate change, environmentaldegradation and social vulnerabilities in Central Vietnam. Weather Clim Soc (4):250–262

Buch-Hansen M (2013) Different ways in decreasing vulnerability to increased climate variability and extremes. Forthcoming. Trans-disciplinary Environmental Studies

Duong B H (2011) 2011-last update, Quang Nam's new integrated culture and tourism strategy for sustainable development in action [Homepage of UNESCO], www.un.org.vn/en/unesco-agencypresscenter [04/04, 2012]

Ensor J (2011) Uncertain futures. Adapting development to a changing climate. Practical Action Publishing, Rugby

Hitchcock M, King VT, Parnwell M (eds) (2009) Tourism in Southeast Asia: challenges and new directions. NIAS Press, Copenhagen

Huq S, Toulmin C (2006) Three eras of climate change [Homepage of IIED]. Available: www.pubs.iied.org/11060IIED.htlm2009]

International Centre for Environmental Management (2006) SEA of the Quang Nam province hydropower plan for Vu Gia Thu Bon River Basin. Tam Ky

IPCC (2007) Climate change 2007, working group II: impacts, adaptation and vulnerability. Fourth assessment report of the intergovernmental panel on climate change. Cambridge University Press, Cambridge

Kelly PM, Tran VL, Hoang MH, Nguyen GHN, Adger WN (2001) Managing environmental change in Vietnam. In: Adger WN, Kelly PM, Nguyen HN (eds) Living with environmental change. Routledge, London

Quang Nam Provincial People's Committee (2010) Annual socio-economic development reports of Quang Nam (2005–2010). Tam Ky

Stern N (2006) What is the Economics of Climate Change? World Economics 7(2)

The Economist (2012) Vietnam. From hero to zero: the communist party sticks to its principles and the economy stalls. The Economist, 31 March 2012 The Library of Congress Country Studies (1987) Vietnam Country Study [Homepage of The Library of Congress Country Studies], http://lcweb2.loc.gov/cgi-bin/query/r?frd/cstdy:@field%28DOCID+vn0071%29 May, 2012

The Library of Congress Country Studies (1987) Vietnam Country Study [Homepage of The Library of Congress Country Studies], http://lcweb2.loc.gov/cgi-bin/query/r?frd/cstdy:@field%28DOCID+vn0071%29 May, 2012.

Vietnam General Statistics Office (2009) I. Statistical Publishing House: Hanoi

Vietnamnet (2012a) Hydropower water reservoirs lend a hand to illegal lumberjacks [Homepage of VietNamNet],.vietnamnet.vn/en/environment/20088/hydropower-water-reservoirs-lend-a-hand-to-illegal-lumberjacks.html [04/04, 2012]

Vietnamnet 03/28/2012 (2012b) Over 40 hydro-power dams in Quang Nam to be checked [Homepage of VietNamNet] vietnamnet.vn/en/society/20512/over-40-hydro-power-dams-in-quang-nam-to-be-checked.html [04/04, 2012]

World Bank (2010a) The Social Dimensions of Adaptation to Climate Change in Vietnam. Discussion Paper Number 17. World Bank, Washington DC

World Bank (2010b) World Development Report, 2010. Development and Climate Change. World Bank, Washingthon DC

Is Climate Change a Reality for Agriculture in Quang Nam Province?

Ngoc Quang Vu, Henning Schroll, Jan Andersen and Søren Lund

Abstract The chapter describes changes in agricultural systems in Quang Nam Province since 1990 in an effort to explore if such changes are reflecting strategies of adaptation to climate change. A two-pronged analytical strategy is followed. This explores (a) patterns of correlation between agricultural production, eco-agricultural landscapes, and climate variations, and (b) records stakeholder perceptions of the causal relations between changes in agricultural practices and climate change. It concludes by discussing the concordance and differences found between the two analyses. An overview of agricultural changes was provided through available statistics and interviews and important climate impacts and adaptations were identified and assessed. An energy flow method was used to describe the agricultural situation. From interviews it was found that many Vietnamese authorities and farmers are convinced that climate change is having an impact on farming, forestry and aquaculture today, and consequently, different kinds of adaptations are being introduced. The study shows that farmers in the agricultural field during the last decade have increased productivity and demonstrated a strong ability to adapt farming systems to the impacts experienced from extreme weather events. Forestry adaptation to extreme weather impacts took a long time while aqua culture adaptation occurred rapidly. With regards to correlations of documented patterns of climate variations and changes of agricultural production systems, however, the findings of the study are inconclusive. Long term planning should include concerns of adaptation to climate change and the use of economic incentives.

Keywords Agricultural production • Climate variations • Aquaculture • Extreme weather • Land use planning

N. Q. Vu
Institute of Geography, Vietnam Academy of Science and Technology, Hanoi, Vietnam

H. Schroll (✉) · J. Andersen · S. Lund
Department of Environmental, Social and Spatial Change, Roskilde University, Roskilde, Denmark
e-mail: schroll@ruc.dk

O. Bruun and T. Casse (eds.), *On the Frontiers of Climate and Environmental Change*, Environmental Science and Engineering, DOI: 10.1007/978-3-642-35804-3_3,

1 Introduction

Vietnam is likely to be among the countries hardest hit by climate change, mainly through rising sea levels and changes in rain fall and temperature. Agriculture could be extensively affected by climate change, and finding effective adaptation strategies to climate change will be critical for maintaining food security (Yu 2010). Model studies for the Southeast Asian region show that climate changes could lower agricultural productivity as much as 15 percent in Vietnam by 2080 if global warming continues unabated (Zhai 2009).

Quang Nam is a province in Mid Vietnam with a vast mountainous area in the West and a narrow coastal plain in the East. Dry hot weather constitutes the climate throughout most of the year with typhoon influence from the East Sea. The cultivation of rice takes up 75 % of the agricultural area and other crops are peanuts, maize and industrial crops like sugarcane, cashew nuts and coconuts on remaining farmland. Agriculture (including farming, forestry and aquaculture) has in recent years experienced a relative decline of economic importance in Quang Nam province. In 2009 agriculture contributed 13 % of GDP (Gross Domestic Product) compared to 36 % for the industrial sector. (Statistical Yearbook 2009).

The purpose of this chapter is to analyze the development of agriculture in Quang Nam. Changes of agricultural systems in Quang Nam are documented and understood in an effort to explore if such changes are reflecting strategies of adaptation to climate variations as perceived by farmers and local authorities, and as compared to scientific data on climate events. Finally, the study will suggest some directions for more long term adaptations of agricultural systems in Quang Nam.

The approach adopted in the study is based on two main underlying premises. As a first premise, we claim that although the loss of agricultural crops, physical infrastructure and human lives are associated with the occurrence of extreme weather events, it should be borne in mind that the nature and seriousness of such losses are also influenced by, and should be interpreted in the light of technological, demographical and socio-economic factors (DFC 2008). This calls for interdisciplinary and interactive models of explanation rather than simple cause-and-effect relations. A second premise lies in the study's intention to inform decision-making on the subject of responses to climate change, in a perspective of *adaptive planning* (Weber 2006), where stakeholder involvement is pivotal in knowledge production and situational analysis. This has prompted a choice of actor-sensitive and participatory research methodologies that aim to develop methods and procedures which can be used for the planning of future responses to climate change within the agricultural sector in Quang Nam, and which subsequently might be replicated in other provinces in Vietnam.

2 Framework and Approach

The study was conducted according to a basic interdisciplinary conceptual framework. This understands impacts of climate change and extreme weather events on agricultural production systems as dependent on (1) the specific nature of local climate change (2) the type of eco-agricultural landscapes and agro-ecological zones (see definitions below) present on the location, and (3) a range of interdependent demographic, institutional, and socio-economic factors given in the local context. In accordance with this framework, adaptation and responses to climate variations should be planned based on local analyses of the specific situation of the various impacts of climate on different types of agro ecosystems with regard to geographical location and socio-economic factors.

Specifically, this has implied (a) exploring patterns of correlation between agricultural production, eco-agricultural landscapes, and climate variations, and (b) recording (or giving voice to) stakeholder perceptions of the causal relations between changes in agricultural practices and climate change, and discussing the concordance and differences found between the two analyses.

To implement this, a stepwise approach was used. *The first step* was composed of two parallel tasks. First, the identification of existing eco-agricultural landscapes present in the province was undertaken. Parallel to this, the concept of climate variations needed to be defined and the identification of climate indicators determined. On this basis, a collection of available statistics and interviews with stakeholders and authorities were conducted in order to get an approximate idea about the local climate change situation. *The second step* included the collection of data on existing land-use and production systems in and modelled through an energy flows analysis *(see explanation below)*. As a *third step,* local perceptions of changes in the agricultural production systems over the last 10–20 years and the causes of these, as well as local perceptions of the underlying causes were collected.

The *term eco-agricultural landscape* in this study is used in the sense given by Scherr (2008) to designate land use mosaics of areas in natural/native habitat and areas under agricultural production. These landscapes are defined as "spatial units needed or actually managed by the group of stakeholders working together to achieve biodiversity, production and livelihood goals" (Ibid). The term is used in combination with a methodology for zoning and analysing agricultural systems. This is in order to plan and prioritise research and development activities in the fields of agriculture and natural resource management, termed *agro-ecosystem analysis* (AIA). AIA operates with the notion of *agro-ecological zones,* which are defined as areas with fairly homogenous biophysical and socio-economic conditions or characteristics (Jones 2005). Eco-agricultural landscapes of land use mosaics are thus grouped together in agro-ecological zones according to selected biophysical and socio-economic characteristics.[1] These are considered as important because they reflect local situations of exposure and vulnerability to climate change.

[1] Working language being English, the Danish members of the team faced some translation problems during interviews with non-English speaking persons. Mistakes under translation were quite common in the beginning of our cooperation. During interviews, problems appeared in relation to formulating correct questions and also to understanding the answers. Some mistakes appeared for example concerning the correct yields of crops because farmers in the area harvest several crops per year so the term "growth seasons" were sometimes confused with the term "years". In order to avoid this, all information was double-checked. As the time went by, the cooperation between the Danish and the Vietnamese research teams improved and data collection activities were more efficient during the second fieldtrip.

2.1 Energy Flow Analysis

To get an overview of farming processes and techniques an energy-flow analysis frame was used. The analysis includes quantification of the production and it increases the ability to assess the importance of climate (lower production) or technological development (higher production). Energy-flow calculations includes energy out-put (contents of energy in vegetable produce, animal production and the part of the produce that was used for human consumption), and energy in-put (which includes the in-put of energy used for the production of fertilizer and pesticides, the energy spent in the physical efforts of human labour, the physical work of water buffaloes, in the production of machines, as well as direct in-puts of fuel and electricity). The ratio of out-put of the production used for human food to energy in-put is called *the out-put/in-put energy ratio*. The ratio will decline in case of an increase in the use of fertilizers or other energy in-puts or if a larger part of the vegetable production (out-put) is used as fodder for animals. The energy in-put from the sun for the photosynthesis is large but it is kept out from the energy ratio calculations, because the focus for the energy account is the fossil energy in-put (Schroll 1994).

The energy-flow method is beneficial to provide an overview of the technological level and quantitative out-puts of the farming process. The energy-flow analysis acts as a frame against which direct and indirect climate variations and adaptation measures can be assessed. In the present study, energy flow analyses were omitted for forestry and aquaculture as the necessary statistical data were not easy to provide.

2.2 Field Work

Among provinces in Mid-Central Vietnam Quang Nam was selected because it contained different eco-agricultural landscapes and there was relative easy access to these landscapes. In order to cover different agricultural systems and different eco-agricultural landscapes three field trips in October 2009, January and November 2010 were carried out by Vietnamese and Danish scientists. Quang Nam province is divided into 18 districts which consist of 210 communes each containing a number of hamlets (villages)—the smallest administrative unit. Before visiting a specific agricultural production unit or area, the authorities at provincial-, district-, commune- and hamlet levels had to give their permission. The visits were planned to cover typical farming, forestry and aquaculture in different landscapes. The selected sites of visits in districts and communes appear from Table 1 (Vu Ngoc Quang 2011).

The agricultural practices were studied by means of field observations during site visits and during interviews with farmers and local authorities. In each of the five visited districts interviews were made with district and commune

Table 1 The districts and communes visited during fieldwork 2010. (Districts appear from Fig. 1)

District	Commune	Landscape ecology for agriculture
Thang Binh	Binh Hai	Coastal sandy plain
Nui Thanh	Tam Hoa	Coastal sandy plain
Dai Loc	Dai An	Alluvial plain
Que Son	No commune visited	Abrasion and accumulation plain
Bac Tra My	Tra Giang and Tra Duong	Eroded plain area in mountains
Nui Thanh	Tam Thanh and Tam Tra	Eroded plain area in mountains

authorities. The semi-structured interviews covered agricultural issues needed for the energy flow analysis, climate change and impacts, and the knowledge and perceptions of the authorities on which changes were considered positive and which ones were negative. Civil servants were present during interviews with the farmers. Danish and Vietnamese researchers worked together in the field and made the interviews. The data was critically triangulated by cross-checking the information provided by the respondents with our own observations and especially with available official statistics. The teams focused on the activities farmers had carried out in order to avoid negative consequences on the agricultural production by climate impacts, and how and from whom the farmers had got the information. Interviews were carried out with experts and authorities at the provincial level. Relevant legislation and reports were collected at the same occasion. An important methodological challenge was to assess the reliability of the statistical data available. In Vietnam, the provincial authorities collect quite detailed statistics on current land uses, farming crops and animals, forestry and aquaculture which are published annually. Ensuring the reliability of the statistical data was an important task. The team visited the statistical office at Quang Nam provincial level and discussed statistics of agricultural products and we carried out the same procedure in district Bac Tra My.

In Vietnam, the rules are that all sellers of planted forest trees need permission from the district. Reports on the status of the planted forest, harvested out-put and diseases and suggestions for the future forest production are sent to Department of Agriculture in the Province every 6 months. Comments, suggestions for using new equipment, and subsidies are subsequently sent back. The district has an annual meeting with the Province Department to discuss possibilities of improving the forest production and they have often administrative meetings with the Province Department.[2] They followed international standards for statistical data collection.

[2] Phan Thanh Phuong 2010 "Interview". Vice Director of the Department of Agriculture and Rural Development. Bac Tra My District. 8.01.2010 and 17.11.2010.

3 Agro Ecosystems in Quang Nam

Quang Nam is located in a tropical zone, characterized by a humid monsoon climate. Annual average temperature is 25 °C and it is hot and dry from February to April and rainy form September to December. The diversified eco-agricultural landscape systems within the province have emerged due to important geo-physical variations. The local agricultural productions systems differ considerably according to geographic locations and the specific socio-economic and physical elements. They are, therefore, affected differently by extreme weather events and rises in sea levels (Vu Ngoc Quang 2011).

The landscape systems can be divided in a number of agro-ecological zones according to altitude and geological origin of the soils. The plains in Quang Nam can be divided into three zones: the coastal zone with aquaculture and farming; the alluvial plain landscape with farming; and the abrasion and accumulation plain landscape, also with farming. Upland agro landscapes are characterized with forestry and a little farming. Agriculture is different in the geographic locations and consequently the climate impacts are also expected to be different.

3.1 Coastal Sandy Plain Landscape

Coastal sandy plains are located right beside alluvial plains or abrasion and accumulation plains, forming sandy ranges along the coast, see Fig. 1 map of Quang Nam (marked with yellow colour). Affected by wind, some sandy dunes are shaped in the central area of the sandy range. The landscape of the sandy plains is flat and normally higher than surrounding alluvial plains. There is easy drainage in this landscape but soil quality and the capacity for keeping water is poor. These plains are not very favourable for farming but suitable for human habitat. Most urban and rural residential areas of Quang Nam are located in the sandy areas. Rice and other crops are grown in many areas of the sandy plains where intensive cultivation techniques are used and irrigation systems can supply enough water for farming. The sandy plains are located next to estuaries by the sea so it is supportive to fishery and aquaculture which provide one of the main income sources for local people. (Vu Ngoc Quang 2011) (See Fig. 1).

3.2 Alluvial Plain Landscape

This landscape is located widely in the lower sections of rivers: Thu Bon, Truong Giang and Tam Ky, and occupy the majority of areas of Dai Loc district (See Fig. 1). Although the landscape accounts for only 1/8 of the total area of Quang Nam, half of the population of the province are living in this landscape because

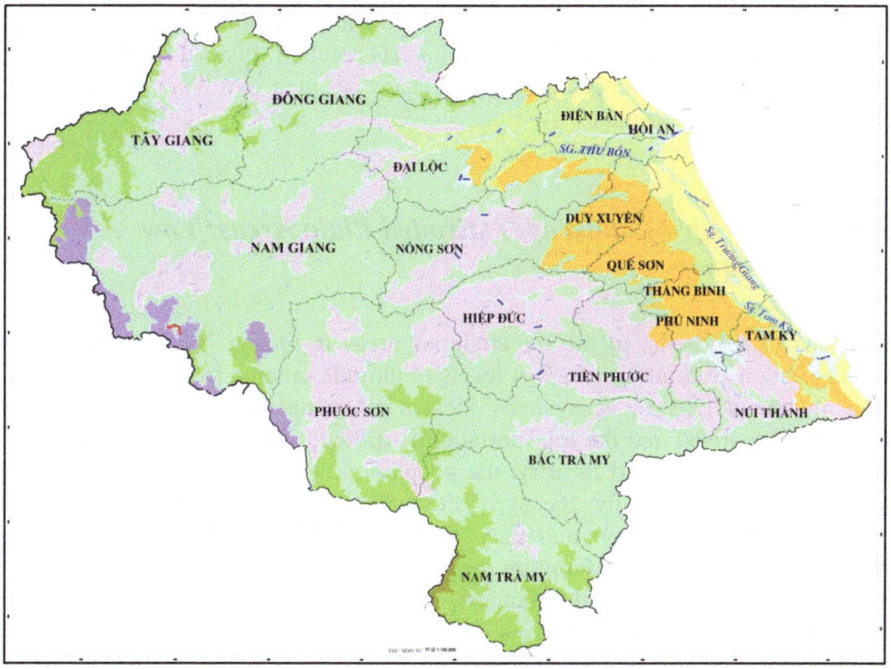

Fig. 1 Map of Quang Nam province showing districts, major rivers and landscapes. Plains are indicated by *yellow*, *yellow green* and *orange*. Low mountains are *light green* and *red blue*. High mountains are *green*, *blue* and *brown* (Vu Ngoc Quang 2011)

it has many advantages. Growing paddy rice combined with breeding livestock and industrial crops are typical features of agro ecosystems in this landscape. (Vu Ngoc Quang 2011) (See Fig. 1: Yellow green colour).

3.3 Abrasion and Accumulation Plain Landscape

This landscape is higher. Abrasion and accumulation plains are located next to highland areas in the west of Quang Nam, belonging to the districts Que Son, Thang Binh, Phu Ninh and Nui Thanh. This landscape consists of lowland areas lying next to low hills. Soil in lowland areas is mixed with sand and gravels with poor quality and used for growing rice and other crops. In the abrasion and accumulation plains, agro ecosystems of paddy farming occupy about 1/3 of the total area. The low hills are used for growing acacia and eucalyptus. Hydrographical systems in these areas cannot supply enough water for drinking and for production. In areas with no irrigational systems and reservoirs, local people are affected by droughts every year. At locations with access to the irrigational systems and reservoirs, where water for production is ensured, a change in the structure of agro ecosystems has resulted. Instead of one rice crop per year, many areas can produce 2 rice crops per year (Vu Ngoc Quang 2011) (See Fig. 1: orange colour).

3.4 Upland Agro Landscapes (Eroded Plain Areas: Low Mountains)

Most agro ecosystems in the mountainous areas in Quang Nam are located in ancient flatten areas along rivers at levels of altitude ranging from 100 to 1,200 m (low mountains). Agro ecosystems located in mountainous areas are relative small and separated and the transportation is difficult. Due to poor conditions for residence, cultivation and forestry, the population density is low, 10–15 persons/km^2 on average. Typically, the agro ecosystems depend on rain water because there is no irrigation system and reservoir. Most areas are used for growing paddy rice and one crop per year. Farming methods and technology comprise low capital and energy in-put and non-mechanized, mainly subsistence farming. The out-put is low in these areas because there are no favourable conditions for intensified cultivation (Vu Ngoc Quang 2011). Even though the areas of the highland districts are large compared to other districts, the yearly rice production is around 10 % of the rice production in districts with alluvial plain landscapes (Statistical Yearbook 2009) (See Fig. 1: light green; red blue).

4 Farming in Quang Nam

The farming and forest activities are grouped together in the official national statistics and GDP is 10 % while industry is almost 4 times as important in economic terms (Statistical Yearbook 2009). Paddy rice is quantitatively the most important crop in Quang Nam province, but the rice production is not technologically as developed as in provinces in North and South Vietnam. Technological in-puts in agricultural production are pesticides, fertilizers and hand driven machines. Rice production is smaller in the province compared to North and South Vietnam, machines are few and buffaloes deliver the main power in the production (Le Moun 2010).

Vietnam has a mixture of a state planning and a free market economy so government plays important role in determining the size and composition of the national agricultural production. Every year, the Provincial Government prepares a plan foreseeing the crop out-put for the coming year, the so-called plan targets. The actual production—for example of rice—is subsequently assessed according to the planned targets. This is what is expressed in the statistical yearbook when it is recorded that the rice cultivating area in the province was 104.7 % more

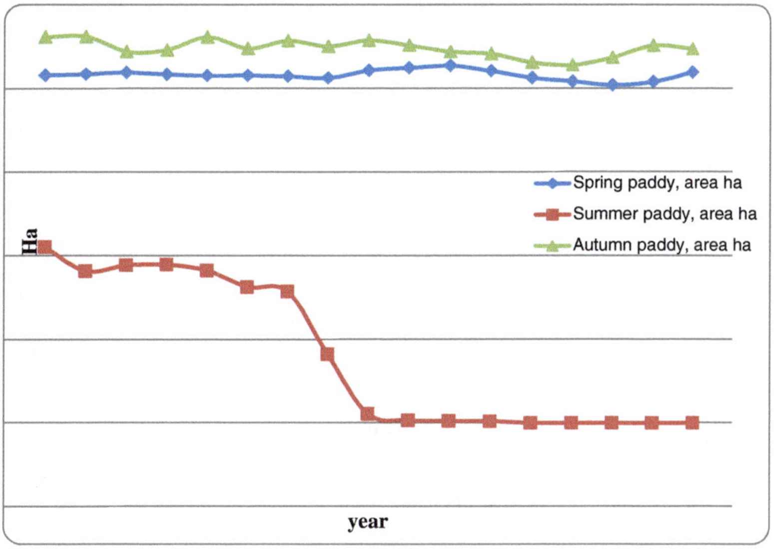

Fig. 2 Rice production, 1993–2009

than expected in the plan and the production of rice equalled to 96.95 % of the plan in 2008 (DARD 2010). In 2009, the total yield of rice was 3,94,000 tonnes including the farmers' own consumption. The plan does not mention fertilizers or pesticides. The local authorities publish a detailed guidance on many practical issues like time for planting and the use of fertilizer. The farmers choose themselves how much fertilizer to be used in their fields and what crops they want to grow. There are more than 20 varieties of rice in the province and the varieties are selected mainly on the basis of different soil quality. In some years, for example 2009, the province has had a surplus of rice for export out of the area (Le Moun 2010).

The agricultural statistics are detailed and cover rice and other vegetable outputs from farming such as maize, sweet potatoes, peanuts and soybeans. For the paddy rice production the statistical reports cover the planted area for three seasons over the years from 1993 to 2009, see Fig. 2 (Statistical Yearbook 2002, 2009).

Figure 2 shows that rice production has changed from 3 harvests to two harvests per year. For 2009 the rice harvested was 3,94,000 tons and the area was approximately 55,000 hectares (Statistical Yearbook 2009).

The Spring and Autumn paddy rice were not grown in exactly the same area from season to season because other crops or fallow land could occupy some of the areas. The total rice production is given in Fig. 3 (Statistical Yearbook 2002, 2009).

For the districts visited the average rice out-put on district level is shown in Table 2. The yields were quite different in the different districts depending on soil fertility (see Chapter 3) and the use of fertilizer and pesticides.

Figure 4 shows an increase in maize and a decrease in sweet potatoes. The cassava area is increasing but part of this production takes place in the planted forest

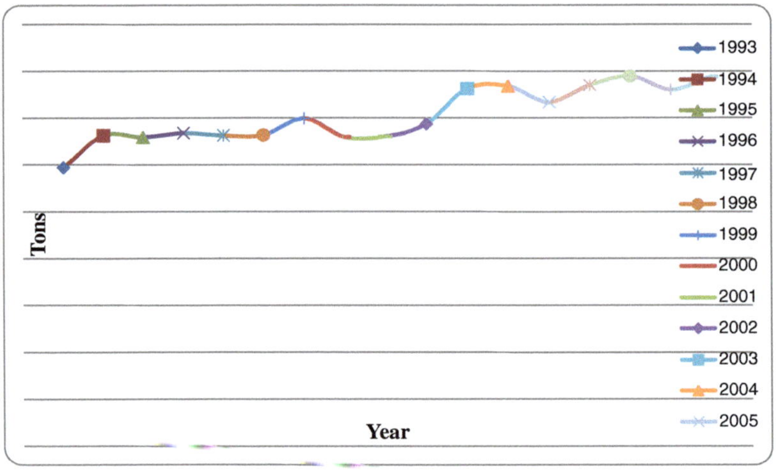

Fig. 3 Total harvest tons rice per year in Quang Nam

Table 2 The yearly spring and autumn paddy rice production in the visited districts was calculated for year 2009

District	Two rice crops in the area[a] ha	Total rice harvest in tons	Tons per ha per year
Thang Binh	6,997	60,941	8.7
Nui Thanh	3,763	31,633	8.4
Dai Loc	4,399	49,145	11.2
Que Son	3,664	32,961	9.0
Bac Tra My	1,037	6,799	6.6

[a] The areas in spring and autumn—paddy rice areas were not the same so an average area was used for each year
Source: Statistical Yearbook 2009

areas.[3] Cassava is grown on sloping soils of the mountainous areas of West Quang Nam and the crop has the ability to grow on poor soils (Global Cassava 2012).

Peanuts and soybeans occupy increasing areas. Coconuts, sugar cane, cashew nuts and tea are declining industrial crops, while rubber is increasing (Statistical Yearbook 2009).

4.1 Energy-Flow in Rice Production

An energy flow analysis has been undertaken of paddy rice production as paddy rice is the most important crop in Quang Nam province. The energy out-put from

[3] Phan Thanh Phuong 2010 "Interview". Vice Director of the Department of Agriculture and Rural Development. Bac Tra My District. 8.01.2010 and 17.11.2010.

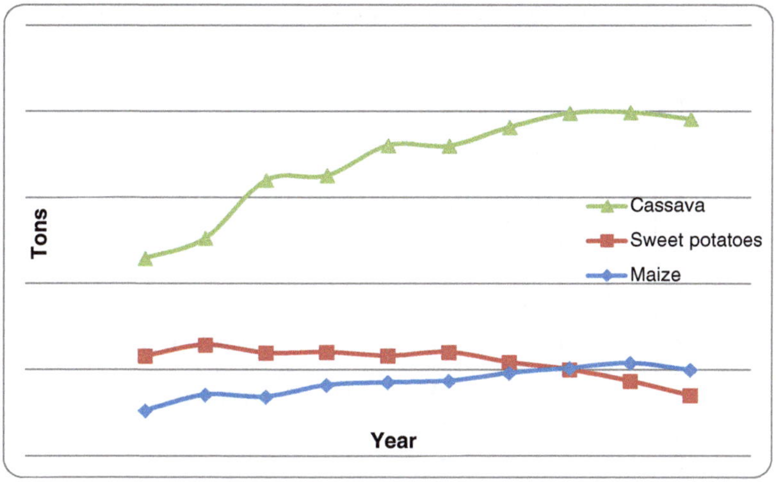

Fig. 4 Other quantitative important farming crops in Quang Nam (Statistical Yearbook 2002, 2008, 2009)

the farms and the energy in-put to the production is calculated and important parameters are shown in Fig. 5. Out-put is rice and meat. We have omitted meat from the calculations and only included ox as working animals. The reason is that rice is mainly used for human food and not for fodder. Energy in-puts in paddy rice production in Quang Nam are fertilizer, pesticides, machines, fuel and electricity, labour and ox power. Energy in-put is the energy used to produce the fertilizers, pesticides, machines etc. (Schroll 1994). Energy out-put is represented by the energy content in the rice harvest.

Information on the specific energy in-puts needed for the analysis cannot be found in the national statistics, so our data on these are based on interviews and consequently figures become uncertain. In Table 3 some data are listed from interviews. Fertilizers are used in the paddy rice production as can be seen from Table 3, where three farmers gave rather different information about their use of fertilizer per ha and crop. A guide for Quang Nam farmers recommended 540 kg fertilizer per ha and crop (Quang Nam Agricultural Department 2009). Since there are two harvest seasons—spring and autumn—we chose to apply a total amount of 1 t of fertilizer per ha per year as the basis of the calculation. On an average there were 43,000 ha paddy rice in Quang Nam in Quang Nam Agricultural Department 2009, (Statistical Yearbook 2009). To produce 1 t of NPK fertilizers energy consumption is equivalent to 0.32 t of oil (Schroll 1994). The energy content of 1 t of oil is 4.2×10^7 k j. For Quang Nam, the consumption of 43,000 t of NPK is equivalent to an energy in-put of 57×10^{10} k j.

Information on the use of pesticides is difficult to obtain. In Table 4, a use of 10 L of active substance per ha was given as an answer by one of the respondents. Spring rice is vulnerable to Rice Blast disease and prophylactic pesticides were

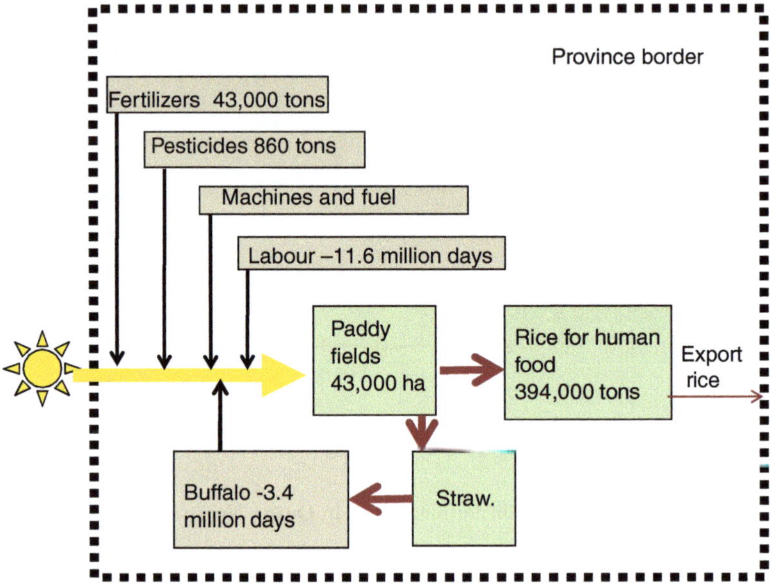

Fig. 5 System Ecological model of paddy rice fields 2009 for Quang Nam province (two crops per year)

Table 3 Data from interviews of paddy rice farmers in mountain area, plain area and coastal area

In-put/out-put to paddy rice for one farm	Farmer Mrs. Vo Thi Quyen. Coastal zone. District Nui Thanh, Commune Tam Hoa	Farmer Mr. Vo Viet Duong. Plain area. District Nui Thanh, Commune Tam Thanh	Farmer Mrs. Nguyen Thi Thuan. Mountain area. District Bac Tra My, Commune Tra Duong
Rice field area in m^2	2000	2000	720
Fertilizer kg NPK per ha per crop	500	800	300
Pesticides—active substance liter per ha per crop	No information	10	No information
Rice production tons per year	1.2	1.6	0.4
Rice production tons per year and ha	6.0	8.0	5.5

recommended (Quang Nam Agricultural Department 2009). From this information we assume the yearly consumption is one spraying of 2 crops per year—in total 86,000 L. The energy consumption to produce one kilo of pesticide is 2,00,000 k j (Schroll 1994). The total energy contribution is 2×10^{10} k j.

Table 4 Energy out-put in-put for Quang Nam paddy rice production

	Type	Unit	Energy 10^{10} kilo joule
Energy in-put	Fertilizer	43,000 tons	57
	Pesticides	860 t	2
	Labour	11.6 million days	14
	Buffalo power	3.4 million days	62
	Machines, fuel	No information	
Energy out-put	Rice	3,94,000 t	591
	Straw	No information	

In Fig. 5, a category of energy in-puts such as machines and fuel is indicated. However, the team was not able to obtain any statistical data on the extent to which machinery is used, but it was found that ploughs and harvesting machines can be seen in the fields. If the energy in-put contributions from machines and fuel were included in the calculation, the energy out-put/in-put ratio would be lower.

The team has made the assessment that a Vietnamese hard working farmer would need 12,200 k j per working day (Heslin 2007). For the planting and harvesting of 1 ha paddy rice per season, 95 working days and 40 days together with a buffalo—in total 135 man days will be used per crop.[4] If this result is chosen as an average, the use of working days can be calculated as 11.6 million working days per year. The annual energy consumption for paddy field workers can then be estimated to be 14.2×10^{10} k j for 2009.

Water buffaloes are common in paddy field work in Quang Nam. The buffalo stock is 80,000 heads and cattle number additional 2,10,000 heads (Statistical Yearbook 2009). 40 days will be needed for one person to work together with a buffalo per one ha and one crop. During one year there will be a need for an in-put of 3.4 million buffalo days if all field work is made by the use of buffaloes. It is assumed that a working buffalo consumes 50 % more than a non-working one and the extra food will be approximately10 kg of straw per day.[5] 1 kg of straw is 18.2 10^3 k J (Schroll 1994). The working buffaloes get extra fodder equivalent to 34,000 t straw or 62×10^{10} k j.

Energy out-put is calculated from production figures taken from the national statistics. Rice straw is not registered in the statistics. In 2009 the out-put of rice in Quang Nam is 3,94,000 t (Statistic Yearbook 2009). 1 k of rice contains 15,000 k joules (Danske Slagterier 2010). Energy out-put of rice is 5.9×10^{12} k j in 2009.

The different calculations are presented in Table 4. The energy ratio can be calculated. For straw, a figure of the harvest (out-put) was not available, and furthermore it should be considered that some straw is used for paper production.[6]

[4] Nguyen Ngoc Nuoi 2010 "Interview". Vice Chairman Nguyen Ngoc Nuoi, Tam Tra Commune, Nui Thanh District 14.1.2010.

[5] Dang Thi Thuy 2010 "Information from interview with 6 Vietnamese farmers from Hanoi. Project secretary, Institute of Geography, Hanoi 20.01.2010.

[6] Nguyen Van Thong 2010 "Interview". Manager of Industry and Trade Department of Quang Nam Province16.11.2010.

Therefore buffalo power has been excluded as an in-put for the calculation of the ratio because working buffaloes mainly feed on straw and grass.

The energy ratio out-put/in-put for Quang Nam 2009 is 8.1 (without buffalo power in-put). Similar calculations made for Danish agriculture in 1990 gave an energy ratio of 1.0 (Schroll 1994).

The Quang Nam farming gives a high yield compared to energy in-put and that is due to a relative small energy in-put and to a situation where most rice is eaten by people. The farming is not yet industrialized and not so dependent on the cost of chemicals and machinery.

4.2 Climate Change and Adaptation in Farming

Local authorities and farmers were interviewed about their perceptions of climatic changes and ensuing challenges for local farming activities. Climate change problems in farming were known to all interviewees, and farmers expressed that they particularly face problems with higher temperatures, less rainfall, longer dry periods and more flooding.

Many interviewees argued that less rain and higher temperatures create droughts in summer. This issue is most pronounced for paddy rice farming in areas with poor soil and where irrigation is insufficient. In Thang Binh district temperatures have increased in recent years, and during the hot season temperatures often reach up to 36–40 °C, which has had a big impact on economic activities and peoples' lives in the district's sandy coastal areas.[7] The high temperatures have reduced the rice production in the dry season, the duration of which has been extended due to climate change. In the dry season, the area of rice production has been reduced by 10–20 % in Thang Binh.[8] The same observation was made in the Que Son district. In 1980 temperatures of 35–36 °C were very rare. But now 37–38 °C are experienced more often.[9] Furthermore, less rain in the growth season has caused extended drought periods in Que Son district and as a consequence it has become difficult to get enough water for farming. Less rainfall has also been observed in Nui Thanh district. Earlier the rainfall was 3,000–3,200 mm per year. In 2010 however, the rainfall had been reduced to 2,800 mm per year. The rain periods have been shorter, more uncertain and more intensive.[10] While the rain period usually comes to a halt in November, in 2008 the rain period lasted until December, which meant that the small rice plants could not grow. In 2009, heavy rain delayed the harvest, which was reduced by 3,000 tonnes of rice.[11]

---

7 Ho. Van Chung 2010 "Interview". Vice-director of Binh Hai commune. Visit. 19.11.2010.

8 Ho. Van Chung 2010 "Interview". Vice-director of Binh Hai commune. Visit. 19.11.2010.

9 Nguyen Cong Hua 2010 "Interview". Manager of Agriculture and Rural Department, QueSon District, People's Committee. 10.1.2010.

10 Nguyen Minh Khai 2010 "Interview". Vice–chairman, District of Nui Thanh 13.1.2010.

11 Nguyen Cong Hua 2010 "Interview". Manager of Agriculture and Rural Department, QueSon District, People's Committee. 10.1.2010.

It is difficult to find evidence for these changes in temperature and rain that have been noticed by local people and authorities, in the meteorological data for the province over the last 5 years (Statistical Yearbook 2009). Temperature and rain patterns vary between the 12 weather stations but no valid trend can be seen in the local data.

Although the perceived climate change cannot be supported by evidence from the official weather data, there still could be an impact on the agricultural productivity. A widespread argument among the interviewees for the claim of an ongoing climate change appearance is that farmers had to reduce the rice seasons from 3 to 2 per year because of drought in summer. Interviewees said that the shift from 3 paddy rice crops to 2 paddy rice crops per year was a solid argument for claiming the existence of an on-going climate change impact.There is not enough water for 3 crops and the numbers of seasons have been changed.[12] However, as highlighted in Fig. 2, the change in crop production occurred rapidly over a two year period, so it is more likely due to socio-economic circumstances than climate change.

In order to adapt to changes in rainfall levels, farmers have furthermore started to switch to more drought-resistant crops and drought resistant rice varieties. However, as these generate a lower yield and most farmers cannot purchase more fertilizers to compensate for this reduction, farmers have to accept the reduced yield.[13] As a replacement for summer rice, farmers grow crops like maize, peanuts, and soybeans that are more drought resistant than rice. The new crops give farmers a possibility to increase crop rotation on their land which can improve soil conditions and prevent diseases. Some plant diseases might be less violent when crop rotation is higher.[14] Another explanation for the change of summer crops can be that the prices of other crops like maize, soybeans, or peanuts make it profitable for farmers to diversify their crops for sale in the market.[15]

A change in the paddy rice technology applied by the farmers might also make it difficult to have three yearly seasons of rice crops. The direct sowing of rice seeds instead of the use of seedlings increases the growing time, because direct seeds in the field need extra time to grow. Direct sowing requires a sowing bed with limited weed so this technology includes herbicides and also needs a longer growth season. The farmer will save labour for planting rice seedlings and the farmer's choice depends on the costs of pesticides and labour. The team visited fields where direct sowing beds were applied but we do not know the extent of this production technology, since pesticide statistical data is not available. The use of pesticides is closely related to a more industrialized agriculture. A higher labour price makes pesticides needed in the field.

[12] Phan Duc Tinh 2010 "Interview". Vice Chairman, Dai Loc district, city of Ai Nghia. 11.1.2010.

[13] Nguyen Cong Hua 2010 "Interview". Manager of Agriculture and Rural Department, QueSon District, People's Committee. 10.1.2010.

[14] Le Moun 2010 "Interview". Manager of Division of Plan and Finance. Agriculture and Rural development Department, Qouang Nam Province. 12.1.2010.

[15] Nguyen Minh Khai 2010 "Interview". Vice–chairman, District of Nui Thanh 13.1.2010.

Vietnamese farmers are free to buy pesticides in shops and they are not informed about any official approval and control procedures.[16] One restriction on the use of pesticides is that 50 days before harvest, the use of pesticides on crops is not allowed.[17] There seems to be awareness about problems related to pesticides because we were told that consumption has been reduced with 30 %.[18] The reason stated was the use of local guidelines and training courses.[19] The team did not succeed in getting any statistical data about the yearly consumption of pesticides or fertilizers in agriculture. Without statistical data on the amount or toxic quality of pesticides, it is difficult to assess a declining consumption.

Local people say that rice production is impacted and reduced by extreme climate impacts. An increased number of serious flooding events are occurring in plains and mountains. In the coastal zones brackish water floods rice fields, which means that the planting/sowing of paddy rice has to be delayed until fresh water has diluted the field and the chosen rice variety can grow. Some rice varieties can grow in higher concentrations of brackish water although salt generates a lower yield.[20]

Although harvests may have been diminished locally and the area used for paddy rice has been reduced by 30 % during the last decade due to the loss of summer rice crops, the statistics show that production has been growing since the 1990s and the productivity of the paddy rice areas has increased, see Fig. 4. In some places like Thang Binh district, the statistics show a fall in yearly rice production for 2008, but in 2009 the production was higher than in 2007 (Statistical Yearbook 2009). Faced with the figures from the statistics of quantitative growth, vice director Ho Van Chung from Thang Binh district argued that the total production of rice during the whole year in the district has not been reduced due to the use of other varieties of rice and the moving of rice fields to less flood exposed areas.[21] The fast compensation for lower rice yields indicates that the farmers quickly adapt to a changed growth situation and they have several response options such as new varieties, fertilizers and replacement of fields.

---

[16] Le Moun 2010 "Interview". Manager of Division of Plan and Finance. Agriculture and Rural development Department, Qouang Nam Province. 12.1.2010.

[17] Phan Duc Tinh 2010 "Interview". Vice Chairman, Dai Loc district, city of Ai Nghia. 11.1.2010.

[18] Phan Duc Tinh 2010 "Interview". Vice Chairman, Dai Loc district, city of Ai Nghia. 11.1.2010.

[19] Pham Van La 2010 "Interview". Vice Chairman. Tam Thanh Commune, Nui Thanh District. 13.1.2010.

[20] Ho. Van Chung 2010 "Interview". Vice-director of Binh Hai commune. Visit. 19.11.2010.

[21] Ho Van Chung 2010"Interview". Vice-director of Binh Hai commune. Visit. 19.11.2010 and Tran Tung 2010"Interview". Vice manager of the Department of Agriculture and Rural Development.Office of Peoples Committee of the District of Thang Binh (Department Administration). 18.11 2010.

5 Forestry in Quang Nam

The forests in Quang Nam cover half of the area of the province and it is divided into natural forests and planted forests, see Fig. 6. According to the vice-director of one of the communes in the province interviewed, natural forests have not been used for commercial purposes and local people have only used wood from planted forests for the last 10 years.[22] This statement from a representative of the local government of Quang Nam does not fit well with the many reports in newspapers describing illegal logging.

Planted forests are present in all districts but the larger areas are situated in the mountains on sometimes very steep areas. In the provincial statistic the category of planted forests is most pronounced in the five districts Que Son, Tien Phuac, Hiep Phuac, Nui Thanh and Dong Giang where 80 % of the area of planted forest in the province is situated. The wood production from these 5 districts amount 73,000 m^3 wood and the other 13 districts deliver annual 100,000 m^3 wood. Summing up, the production of wood in Quang Nam comes from planted forest and in the period 2005–2009 the production was between 150,000–200,000 m^3 per year while firewood amounted to around 5,00,000 m^3 per year (Statistical Yearbook 2009). The province plans to plant 400,000–500,000 ha by 2020. The province, however, needs funding for the implementation of this plan.[23]

In Quang Nam, 66 plants produce timber and furniture for Vietnam and for export to Japan and Europe.[24] Some furniture factories use imported timber from Laos and other factories try to cover their needs by local wood production in special forests. Here, the trees are managed in a way which allows them to grow to a bigger size to suit industrial demand. One company uses logs from planted forest (acacia wood of a diameter more than 30 cm) and the products (chairs and tables) are sold to IKEA that regularly controls the production of the company (FOREXCO 2010). There is currently no primary paper production in Quang Nam, although the Quang Nam government seeks to invest in a paper factory in the future. Acacia wood chops are therefore exported to Japan for paper production.[25] Before 1975, forest production in Vietnam was the responsibility of the government but since 1992 the private sector can buy forests. According to interviews, all wood used for commercial purposes originates from planted forest owned by private people. The Quang Nam government has the power to change the land-use for other purposes. To some extent,

[22] Le Minh Hung 2010 "Interview". Vice-director of Division of Planning and Finance (responsible for forestry), Quang Nam Province Department of Agriculture & Rural Development (DARD) 16.11.2010.

[23] Nguyen Van Thong 2010 "Interview". Manager of Industry and Trade Department of Quang Nam Province16.11.2010.

[24] Nguyen Van Thong 2010 "Interview". Manager of Industry and Trade Department of Quang Nam Province16.11.2010.

[25] Nguyen Van Thong 2010 "Interview". Manager of Industry and Trade Department of Quang Nam Province16.11.2010.

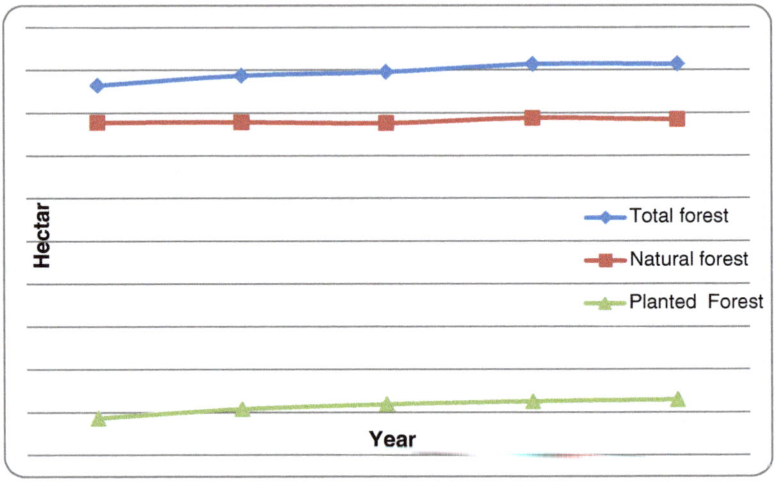

Fig. 6 Total area of forests by type of natural forest and planted forest (Statistical Yearbook 2009)

upland rice fields are being replaced with forests.[26] Many farmers have planted forests in recent years (see "The Push for Plantations: Drivers, Rationales and Social Vulnerability in Quang Nam province, Vietnam" and "Interaction Cultural and Environmental Change: The Co (Cua) Minority of Central Vietnam"). It is easy to expand planted forest because there is plenty of land where forest can be planted. Conversion from rice crops to forestry happens if local people think that it will be economically relevant.[27] Planted forest is not found on the steepest mountains and is always located close to a road in order to facilitate transportation. The transportation of trunks from the mountains to the coast is carried out by private companies.

The concept of mixed forests is defined as forestry systems where at least two species are planted in a forest—one is often Acacia.[28] Acacia is a favoured species as it grows fast and the wood is very useful for producing paper. Acacia has a wide root system that prevents erosion and the wood fibres give strong paper. When the trees grow older and bigger, the logs can also be used for production of furniture.[29] Three types of acacia are used for planted forest: Acacia mangium, Acacia auriculiformis and hybrid acacia.

---

[26] Le Minh Hung 2010 "Interview". Vice-director of Division of Planning and Finance (responsible for forestry) Quang Nam Province Department of Agriculture & Rural Development (DARD) 16.11.2010.

[27] Phan Thanh Phuong 2010 "Interview". Vice Director of the Department of Agriculture and Rural Development. Bac Tra My District. 8.01.2010 and 17.11.2010.

[28] Nguyen Ngoc Bich 2010 "Interview". Tra Giang Commune. 17.11.2010.

[29] Phan Thanh Phuong 2010 "Interview". Vice Director of Department of Agriculture and Rural Development. Bac Tra My District. 8.01.2010 and 17.11.2010.

The Quang Nam forestry sector faces a number of challenges. Acacia species are generally propagated by seeds but Acacia hybrids have a limited capacity for producing seeds so mass vegetative propagation is the only means of multiplication (Galiana 2003). Although the wide root system of the Acacia prevents erosion, soil erosion is difficult to avoid because the forest owners have to cut all trees at the same time due to economic reasons. When all trees are removed there is a high risk of erosion. To reduce erosion a technique of thinning out larger trees can be applied but is often not possible due to forest density (Phan Thanh Phuong 2010).

5.1 Climate Change and Adaptation in Forestry

Typhoons and erosion appeared as key challenges for the interviewees. The typhoons in 2006 and 2009 destroyed 10,000 ha of planted forest in Quang Nam Province.[30] Mr Le Minh Hung from the Quang Nam Province Department of Agriculture and Rural Development has during 25 years of working in the field not seen as serious forest damage as caused by the Ketsana typhoon in 2009. The Ketsana typhoon ravaged the mountainous areas of BacTra My for 3 h, destroying 1,000 acacia trees out of 40,000.[31] Under the same typhoon, 600 ha of the rubber trees were partly damaged, and 345 ha were totally damaged.[32]

As a matter of triangulation, the strong 2009 typhoon is expected to be reflected in the official statistics of wood production. Indeed, a decline of approximately 35,000 m^3 is seen compared to 2008 timber production. However, according to Vice director Mr. Phan Thanh, the decrease of wood production observed was caused by the price of timber and not a direct result of the typhoon in 2009. When the price is high more timber is cut and vice versa.[33] The typhoons break Acacia trees but it is often too time consuming and problematic to cut the broken trees and remove them from dense forest. Although the broken trees can be exploited commercially, the market price of wood will be low when there is a high supply.[34]

To adapt to typhoons a World Bank program has been launched in three provinces of Vietnam, giving farmers economic possibilities for delaying tree felling until they are 7–10 years. After 7 years when the trees are cut a new production

[30] Le Minh Hung 2010 "Interview". Vice-director of Division of Planning and Finance (responsible for forestry), Quang Nam Province Department of Agriculture & Rural Development (DARD) 16.11.2010.

[31] Ha Thanh Ha 2010 "Interview". Farmer in Tra Giang Commune, Bac Tra My province, 8 01.2010.

[32] Pham Van La 2010 "Interview". Vice Chairman. Tam Thanh Commune, Nui Thanh District. 13.1.2010.

[33] Phan Thanh Phuong 2010 "Interview". Vice Director of the Department of Agriculture and Rural Development. Bac Tra My District. 8.01.2010 and 17.11.2010.

[34] Phan Thanh Phuong 2010 "Interview". Vice Director of the Department of Agriculture and Rural Development. Bac Tra My District. 8.01.2010 and 17.11.2010.

can start. All trees are cut in one area at the same time and as mentioned above this exposes the land to erosion.[35]

After the strong typhoons wind sensitive Acacia species have been replaced with the stronger Acacia mangium. This will take 5–7 years and illustrates that adaptation in forestry takes a long time. That hybrid Acacia is sensitive to strong wind has also been experienced in Bac Tra My district, where foresters have started to plant Acacia mangium as adaptation following World Bank funding. Each farmer can get a loan of 10 million dong, which has to be paid back within a period of 3–4 years. The money is used for seed, fertilizer, labours and for the farmers living condition.[36]

In order to convert to forestry or to change the species grown, farmers have to invest a significant amount of money over a long period of time. Currently, there is a World Bank program that supports them. Since 2004 an interviewed forest farmer has planted trees in different areas of his 9 ha in order to have a yearly income. The World Bank program made it possible for the farmer to plant all his land and harvest trees with the same age. During the typhoon in 2009, 30 % of the farmer's trees were damaged.[37]

The standing crop of planted wood is large in the province so the negative impacts of typhoons can easily be compensated for by increasing the cutting of trees. A number of adaptations such as a shift to other Acacia varieties and cutting technique and mixed forests are on-going activities.

6 Aquaculture in the Coastal Zone

In 2009 aquaculture made up 3 % of the GDP in Quang Nam. Out of 18 districts, 6 districts (Tam Ky, Dien Ban, Du Xuyen, Thang Binh and Nui Thanh) have brackish water aquaculture and more than 90 % of the total production came from these districts (Statistical Yearbook 2009). The capacity of fishery and aquaculture has increased because of new technology. Aquaculture production takes place in open ponds with small dikes or in plastic basins in sand dunes. The same ponds are used for production of fish, crab and shrimp at different times of the year. Different types of organisms are never produced at the same time in the same pond.[38] In 2009 the area of fish ponds was 2,250 ha and shrimp aquaculture counted 1,880 ha, both with a falling trend since 2005. The area for mixed and other aquatic products has grown from 5 ha to 312 ha in the period 2005–2009 (Statistical Yearbook 2009). The size of the shrimp ponds is 3,000–3,500 m^2 per

[35] Nguyen Ngoc Bich 2010 "Interview". Tra Giang Commune. 17.11.2010.

[36] Nguyen Ngoc Bich 2010 "Interview". Tra Giang Commune. 17.11.2010.

[37] Nguyen Son 2010 "Interview". Village 3, Commune Tra Duong. Bac Tra My. 17.11.2010.

[38] Nguyen Dinh Son 2010 "Interview and visiting shrimp-, fish- and crab production". Vice manager of Department of Agriculture and Rural Development, 20.11.2010.

household[39] and on average they produce 6 tons shrimps/ha and year (Statistical Yearbook 2009). Mussels are cultivated in the rivers. Small juvenile mussels are taken from the river and placed on certain spots on the bottom where they grow, and after 5 months they are ready for harvesting and sold on the market. Small crabs are taken from the river and raised in special fenced ponds (crabs can walk on land). After 4 months they are ready for harvest. The crab production takes place twice a year.[40]

In Nui Thanh district the area of shrimp production was 1,600 ha in 2005 and 1,500 ha in 2009, so the area was slightly reduced, but the volume of shrimp production in metric tons increased due to new species and better equipment.[41] The same trend of growing shrimp production can be seen from Fig. 7 in recent years, while fish farm production has stagnated. Aquaculture is strongly dependent on fish meat powder which is food produced from small pelagic fish caught in the sea. It is also dependent on sources of seeds of fish, shrimps, lobsters, crabs etc. which were caught from nature and raised in ponds (Nguyen Kiem Son 2010).

Shrimp production is common in Thang Binh in the higher situated sand dunes close to the sea. The visited farmer owned two shrimp ponds of 2,500 m^2 each. The shrimp fry and the fodder are bought from private companies. The farmer pumps water directly from the ocean to replace evaporation and polluted water. A special pond is supposed to store the waste water before treatment. At the time of the field visit, waste water treatment was not working so wastewater was discharged directly from the beach to the ocean.[42] Thang Binh district has difficulty finding solutions to the environmental problems of shrimp production, but they are trying to clean the waste water.[43] Adaptation to the safer higher dune area creates a potential conflict with the tourist industry that prefers clean beaches and water.

The production time for harvesting shrimp is 3 months and after harvest the pond is cleaned. The farmer can use faster growing species of shrimps. From January to July the farmers can produce shrimps. From July to December it is not possible to have any shrimp production in the river. There are two relative low dikes along the river that protect the river banks and the shrimp farms that are

[39] Tran Khanh 2010 "Interview". Manager of Administrative Department. District office in Nui Thanh. 20.11.2010.

[40] Nguyen Dinh Son 2010 "Interview and visiting shrimp-, fish- and crab production". Vice manager of Department of Agriculture and Rural Development, 20.11.2010.

[41] Tran Khanh 2010 "Interview". Manager of Administrative Department. District office in Nui Thanh. 20.11.2010.

[42] Ho Van Pan 2010 "Interview and visit to a shrimp farm in a sandy area near the coast". Village (Hamlet), Dong Tri Village, in Binh Hai Commune.19.11.2010.

[43] Tran Tung 2010"Interview". Vice manager of the Department of Agriculture and Rural Development.Office of Peoples Committee of the District of Thang Binh (Department Administration). 18.11 2010.

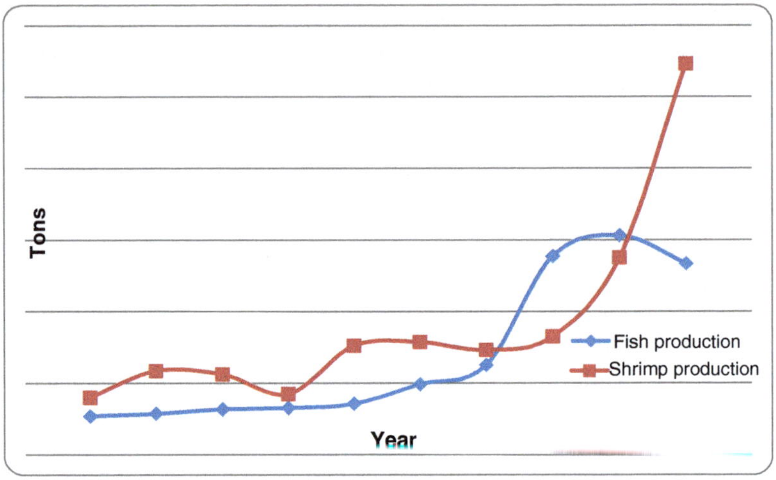

Fig. 7 Production of aquaculture by type of aquatic product during the last decade (Statistical Yearbook 2009)

situated up to these dikes. During the flooding period shrimp production is not possible as shrimp can escape to the river. In recent years the flooding has been so violent that damage to houses and roads has appeared.[44]

6.1 Climate Change and Adaptation in Aquaculture

The interviewees mainly focussed on issues related to typhoons, flooding and warning. In the coastal zone the flooding has impacted aquaculture seriously and caused major changes in technology. Flooded fishponds create a major loss to the farmers who have reacted by introducing a shorter production period to avoid the typhoon season and have tried to find safer localities for ponds in the coastal area. The fish production has stagnated because many fish need a production time of more than 6 months before they can be sold. In contrast some types of shrimps can be produced for market in three months, which makes it possible to farm two shrimp crops per year and avoid the typhoon season. A supplementary strategy has been to move shrimp ponds up in the higher situated sand dunes close to the sea. Figure 7 shows that the growth of shrimp production is high while the aqua fish culture production had stagnated. The area of shrimp ponds has decreased and the production of shrimps has doubled during 2008–2009.

The Nui Thanh District authorities have suggested local people to harvest the shrimps no later than 15 September, when the flooding period starts. There is no

[44] Le Dinh, Thung 2010 "Interview". Farmer in Dong Tri Hamlet in Binh Hai Commune. 19.11.2010.

shrimp production from September to November due to colder weather, typhoons, and the farmers' need to clean the ponds. It is an adaptation to the typhoon season that the shrimp production takes place during 2 seasons. The typhoons and flooding have forced the farmers to produce faster and move their ponds. Shrimp farms have been moved to the sand dunes where the ponds are protected to some degrees from the storms.[45]

Removal of mangrove forests has increased the impact on the coastal zone and aquaculture. Earlier, Thang Binh had a mangrove forest, but this area has now been transformed to shrimp production. Ten years ago mangrove trees were growing in Nui Thanh district area and the mangrove trees created protection from the flooding. But now the mangrove trees have been removed for building of aquaculture ponds. The government wants to launch a program for planting mangrove trees again, but the local people are against this program, because it will reduce their income from the fish-, crab- and/or shrimp production.[46]

Flooding in the coastal zone and the plains is a disaster for households with aquaculture. Farmers have received help from districts and communes and the authorities have established a new unit for repairing flood damage.[47] Weather forecasts communicated by TV and the internet are important sources of information for the inhabitants concerning flooding.[48]

A point of criticism raised many times during the interviews was subsidies. Subsidies were given from the government to local people to reduce the costs of the damages from the flooding. But even though the cost of the damages of the flooding caused by Ketsana was 650 million Dong, only 10 million Dong was given in support. Insurance systems that could cover the losses are non-existent.[49]

The adaptation of aquaculture has been to grow shrimps that can be harvested before the stormy season. This can be identified as a response to the more frequent and more violent character of typhoons and flooding events of the later years.

Still, the typhoons and flooding have forced the farmers to produce differently and move their ponds. Some shrimp farms have been moved to the sand dunes where the ponds are protected in some degrees from the storms. These areas are also important in the tourism industry so there are potential conflicts to be managed.

---

[45] Tran Tung 2010 "Interview". Vice manager of the Department of Agriculture and Rural Development. Office of Peoples Committee of the District of Thang Binh (Department Administration). 18.11 2010.

[46] Nguyen Dinh Son 2010 "Interview and visiting shrimp-, fish- and crab production". Vice manager of Department of Agriculture and Rural Development, 20.11.2010.

[47] Tran Tung 2010 "Interview". Vice manager of the Department of Agriculture and Rural Development.Office of Peoples Committee of the District of Thang Binh (Department Administration). 18.11 2010.

[48] Phan Duc Tinh 2010 "Interview". Vice Chairman, Dai Loc district, city of Ai Nghia. 11.1.2010.

[49] Phan Duc Tinh 2010 "Interview". Vice Chairman, Dai Loc district, city of Ai Nghia. 11.1.2010.

Table 5 Number and intensity of typhoons in Quang Nam during the years (Nguyen Dang Mau 2011)

Years	Level 7	Level 8	Level 9	Level 10	Level 11	Level 12	Level 13
2003	0	0	0	0	0	0	0
2004	**1**	**2**	0	0	**1**	0	0
2005	**2**	0	0	0	0	0	0
2006	0	0	0	0	0	**1**	0
2007	**1**	1	0	0	0	0	0
2008	0	0	0	0	0	0	0
2009	**1**	0	**1**	0	0	0	**1**
2010	**1**	0	0	0	0	0	0

There is no doubt that the typhoons create serious damage for the coastal area. This raises the question of whether typhoons are an indicator of climate change. The Center for Meteorology and Climatology of Vietnam has gathered information on the intensity of typhoons in Quang Nam province which is given in Table 5. (Nguyen Dang Mau 2011).

Table 5 shows that Quang Nam is regularly hit by strong typhoons. Seven years information is to short a period to draw valid conclusion about changes in the intensity of typhoons in Quang Nam as an indicator of climate change. But still, these figures show no changes in intensity during this period. The intensive typhoons in 2006 and 2009 had a devastating impact and affected many people. But it cannot be concluded from the data that climate change was a cause. Besides the time period, Quang Nam is a relatively small area and it is difficult to establish criteria on whether a typhoon that hits part of Quang Nam should be included in Table 5 or not. Conversely, it cannot be argued that climate change has not occurred. The existing data is insufficient to scientifically document that typhoons have become more frequent and stronger. However, although climate change has not been scientifically proven yet, many Vietnamese people engaged in agriculture are convinced that major climate change damage already occurs and adaptations are very urgently needed.

7 Conclusion

Interviewees have unanimously talked about more typhoons, higher temperatures, flooding, erosion, salt water intrusion, heavy rain and serious drought and they have also given examples of specific adaptations made to the perceived climate changes.

The agricultural production in Quang Nam has been statistically described. The study shows that farming has changed from three paddy rice crops to two paddy rice crops per year. The reduction from three to two crops a year happened quite rapidly around 10 years ago (at least in some areas clearly due to local government recommendations; see "Natural Resource Management Impact on Vulnerability in Relation to Climate Change. A Case in Micro-Scale Vietnamese Context"). Some argue that it was due to climate change causing more droughts in summertime and others argue that it was entirely for economic reasons. However, the change happened during just a few years and it took place synchronously amongst all farmers. Within a short time span, all farmers took up both new production and crop rotation systems. This tends to indicate that an effective agricultural information system is in place. In general, statistics show that the output of farming in Quang Nam is growing, while at the same time these statistics do not bear evidence to climate-related reduction of farm yields. The rice growing area was reduced, but production has increased because of new varieties and intensified use of fertilizers and pesticides. The farmers adjust to a new situation whether it is created by market conditions, changes of climate or both. The energy flow analysis indicates that the visited farms are still little industrialised, but we assess that will change in the future. For future adaptation efforts more comprehensive and detailed farming statistics, including data on fertilizers, pesticides, machines and fuel, are essential.

Concerning the forestry adaptation, there is a change to a more robust Acacia species. For further adaptation, better education and enforcement of rules for planting and harvesting may contribute to a more diverse forest management and a reduction of erosion. The World Bank project contributes to forest farmers' adaptation to typhoons.

It is obvious that aquaculture located in coastal areas is seriously impacted by typhoons and flooding; adaptation by means of faster growing shrimp species, higher altitude locations and abstaining from production during the typhoon season has happened fast. Despite flooding problems, shrimp production has doubled from 2008 to 2009. On the other hand it also shows a lack of infrastructural planning, since sand dunes become occupied by shrimp farms, while the beach and sea become polluted with waste water from the shrimp ponds. This industry is potentially in conflict with a growing tourism and more effective coastal zone planning is obviously needed.

Although the prevalence of climatic changes could not been scientifically proven in this study, many Vietnamese people engaged in agriculture are convinced that major climate changes are already happening and require urgent adaptations. The climate events addressed in this study can alert to the potential future impacts of climate change. Recurrent extreme weather events in particular significantly impact on agricultural production in Quang Nam province. While findings from the study show that the Quang Nam population learns and adapts fast to new situations—whether these are spurred by economic or climatic changes—long term adaptation mechanisms to potential climate changes need to be further developed, particular in relation to physical and spatial land-use planning.

Physical and spatial land-use planning will be of value in paddy rice field areas flooded with sea water, because the salty water will hamper a high rice yield. Instead such areas may be used for aquaculture. Planning is an option for restoring mangrove forests in estuaries where trees can protect the coast and agriculture. In the longer perspective careful multidisciplinary planning is needed. Settlements and infrastructure should be localized at places with small risks of flooding and improved possibilities to reach safe places. Presently people do not want to move. In the coming years most households will become richer and have more valuable possessions and they will not accept regular flooding. The compensations would need to be higher. In this situation an insurance system for losses during natural disasters should be required for all households, and the fee for such may be set according to flooding risks in each location. Higher insurance should be meted out for higher risk. By such valuing, insurance may be an instrument to encourage people to settle and produce in a safer manner.

References

Danske Slagterier (2010) www.Danskeslagterier.dk

DARD (2010) DARD department of agriculture and rural development, 2008. Report—Sum-up agricultural production. (Translated by Dang Thi Thuy for sub-project 8)

Danida fellowship centre (DFC) (2008) Application for development research project to the consultative research committee for development research (FFU)—Ministry of foreign affairs. DFC Research Unit, Denmark

Forest Product Export Joint Stock Company of Quang Nam (FOREXCO) (2010) Hamlet 5, Dien Ngoc Commune, Dien Ban district. http://www.forexco.com.vn

Galiana A et al (2003) Micro propagation of Acacia mangium x A. auriculiformis hybrids in Sabah. Bois et Forêts des Tropiques 275

Heslin J et al (2007) Understanding calories, Simon and Schuster, New York

Hoang K et al (1999) Status of cassava in vietnam: implications for future research and development. Ministry of agriculture and rural development, Vietnam. http://globalcassavastrategy.net/Asia/vietnam/v0000e00.htm

Jones P (2005) Agroecosystems analysis, improving livelihoods in the uplands of the Lao PDR. NAFPRI

Nguyen D (2011) Center for meteorology and climatology (CMETC), Vietnam institute of meteorology, hydrology and environment. Ministry of Agriculture and rural development. Hanoi, Vietnam

Nguyen KS (2010) Develop an action plan for aquaculture in line with program of adaption to global climate change in agricultural sector of Vietnam, Ministry of Agriculture and Rural Development. Department of Aquaculture. Hanoi

Quang nam agricultural department (2009) Technical guidance for farmers (Translated by Dang Thi Thuy for sub-project 8, 2010)

Scherr S et al (2008) Biodiversity conservation and agricultural sustainability: towards a new paradigm of "ecoagriculture" landscapes. Philosophical transactions of the royal society B, p 477–494. Royal Publishing, London

Schroll H (1994) Energy-flow and ecological sustainability in Danish Agriculture. J Agric Ecosyst Environ 51:301–310

Statistical Yearbook (2002) Quang Nam Statistical Office. Statistical Publishing House, Beijing

Statistical Yearbook (2008) Quang Nam Statistical Office. Statistical Publishing House, Beijing

Statistical Yearbook (2009) Quang Nam Statistical Office. Statistical Publishing House, Beijing

Vu NQ (2011)In: Nguyen Tran Cau et al. Study on the impacts of climate change on different eco-agricultural landscapes in the study area. Subproject 8: Assessing the impacts of climate change on physical, environmental conditions and socio-economic development in Mid-central Vietnam. Institute of Geography, Vietnam Academy of Science and Technology. Hanoi, p 45

Weber KM (2006) Foresight and adaptive planning as complementary elements in anticipatory policy-making: a conceptual and methodological approach. In: Voss JP et al (eds) Reflexive governance for sustainable development. Edward Elgar Publishing Ltd, Cheltenham

Yu B et al (2010) Impacts of climate change on agriculture and policy options for adaption. The case of Vietnam. International Food Policy Research Institute (IFPRI)

Zhai F et al (2009) Agricultural impact of climate change: a general equilibrium analysis with special reference to Southeast Asia. Manila, Asian Development Bank, Asian Development Bank Institute Working Paper Series 131

List of Interviews

Dang Thi Thuy 2010. Information from interview with 6 Vietnamese farmers from Hanoi. Project secretary, Institute of Geography, Hanoi 20.01.2010
Ha Thanh Ha 2010. "Interview". Farmer in Tra Giang Commune, Bac Tra My province, 8 01.2010
Ho Van Chung 2010. "Interview". Vice-director of Binh Hai commune. Visit. 19.11.2010
Ho Van Pan 2010. "Interview and visit to a shrimp farm in a sandy area near the coast". Village (Hamlet), Dong Tri Village, in Binh Hai Commune.19.11.2010
Le Dinh Thung 2010. "Interview". Farmer in Dong Tri Hamlet in Binh Hai Commune. 19.11.2010
Le Minh Hung 2010. "Interview". Vice-director of Division of Planning and Finance (responsible for forestry), Quang Nam Province Department of Agriculture and Rural Development (DARD) 16.11.2010
Le Moun 2010. "Interview". Manager of Division of Plan and Finance. Agriculture and Rural development Department, Qouang Nam Province. 12.1.2010
Nguyen Cong Hua 2010. "Interview". Manager of Agriculture and Rural Department, QueSon District, People's Committee. 10.1.2010
Nguyen Dinh Son 2010. "Interview and visiting shrimp-, fish- and crab production". Vice manager of Department of Agriculture and Rural Development, 20.11.2010
Nguyen Minh Khai 2010. "Interview".Vice–chairman, District of Nui Thanh 13.1.2010
Nguyen Ngoc Bich 2010. "Interview". Tra Giang Commune. 17.11.2010
Nguyen Ngoc Nuoi 2010. "Interview". Vice Chairman Nguyen Ngoc Nuoi, Tam Tra Commune, Nui Thanh District 14.1.2010
Nguyen Son 2010. "Interview". Village 3, Commune Tra Duong. Bac Tra My. 17.11.2010
Nguyen Van Thong 2010. "Interview". Manager of Industry and Trade Department of Quang Nam Province 16.11.2010
Pham Van La 2010. "Interview". Vice Chairman. Tam Thanh Commune, Nui Thanh District. 13.1.2010
Phan Duc Tinh 2010. "Interview". Vice Chairman, Dai Loc district, city of Ai Nghia. 11.1.2010
Phan Thanh Phuong 2010. "Interview". Vice Director of the Department of Agriculture and Rural Development. Bac Tra My District. 8.01.2010 and 17.11.2010
Phan Van Danh 2010. "Interview". Centre for Agriculture Development Meeting in Quang Nam Province Department of Agriculture and Rural Development (DARD) 16.11.2010
Tran Tung 2010. "Interview". Vice manager of the Department of Agriculture and Rural Development. Office of Peoples Committee of the District of Thang Binh (Department Administration). 18.11 2010
Tran Khanh 2010. "Interview". Manager of Administrative Department. District office in Nui Thanh. 20.11.2010
Vo Thanh Lac 2010. "Interview". Chairman of the Commune, Tam Hoa Commune, Nui Thanh district. 13.1.2010

The Push for Plantations: Drivers, Rationales and Social Vulnerability in Quang Nam Province, Vietnam

Andreas Waaben Thulstrup, Thorkil Casse and Thomas Theis Nielsen

Abstract This chapter compares the impact of land use changes in two municipalities. We look at the long-term changes regarding the expansion of acacia production and the effects of tropical storms. Due to households experiencing significant economic gains after only a few years, both primary forest and agricultural land are being replaced with acacia tree plantations. The downside to this is an increasing social inequality, which follows in the wake of both government decree support given to monoculture promotion, and the influences of market forces. We observe signs of social differentiation, where poor households end up serving as casual labour for the richer families on their acacia plantations. In addition, the poor can be rendered even more vulnerable after becoming labourers for the richer families, because they may no longer have an alternative source of income, yet they still face the risk of an increasing frequency of typhoon exposure.

Keywords Acacia plantations • Destruction of natural forest • Household rationality • Natural disasters

1 Forestry sector in Vietnam

Vietnam's forest cover and the Vietnamese government's forest policies have undergone substantial changes since the end of the Second World War. From 1945 to the mid-1990s, a rapid deforestation process took place, primarily due to widespread logging of the natural forest and land conversion to agriculture. For over a century, upland areas in Vietnam have served as the main source of raw material

A. W. Thulstrup
FAO, Italy

T. Casse (✉)
Department of Society and Globalisation, Roskilde University, Roskilde, Denmark
e-mail: casse@ruc.dk

T. T. Nielsen
Department of Environmental, Social and Spatial Change, Roskilde University, Denmark

O. Bruun and T. Casse (eds.), *On the Frontiers of Climate and Environmental Change*, Environmental Science and Engineering, DOI: 10.1007/978-3-642-35804-3_4,

for the commercial timber industry (Poffenberger 1998). From 1976–1990 contracts were authorised to log 98,000 ha of forest annually (Jong et al. 2006). In the 1980s, State Forest Enterprises controlled over 4 million hectares of forest land which they heavily exploited for timber production purposes, logging 150,000 hectares each year (Poffenberger 1998). These SFEs have been identified as important contributors to the decline in Vietnam's tropical forests due to their logging and milling operations in predominantly upland watershed areas (Jong et al. 2006; Poffenberger 1998). Ngyuen et al. (2010) suggest a number of explanations for rapid deforestation in Vietnam, among the most important ones were: misguided policies, lack of local participation in decision-making and lack of enforcement, especially during the collectivization period, which arguably lead from a de jure state property rule to a de facto open access state.

The overall objective of this chapter is to demonstrate the impact of government strategies to increase forest production on households at the commune level[1] in Quảng Nam province in central Vietnam. We focus on the rationale and decision-making processes of households that have shifted their strategies toward acacia production as well as the socio-economic impact on livelihoods in two communes in Quảng Nam province Forest changes are shown in Map. 1.

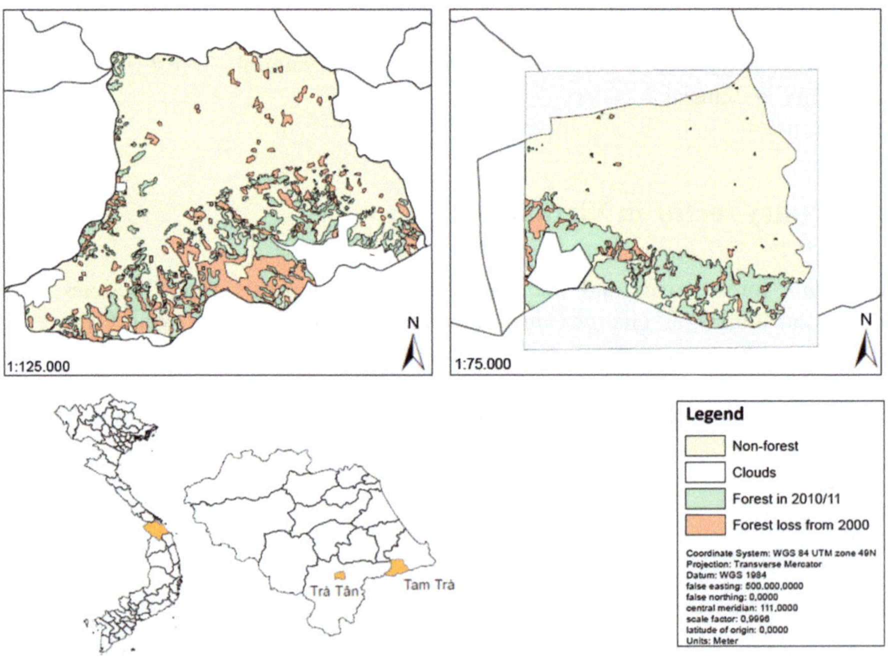

Map 1 Forest changes in Trà Tân and Tam Trà, 2000–2010/2011

[1] The commune (xã) is the lowest administrative level of government in Vietnam.

According to official sources (Table 1), the plantation area increased rapidly by 150 % from 1995 to 2002 while somewhat surprisingly the natural forest cover also expanded, albeit less dramatically, by 25 %, from 8.3 to 10.4 million hectares. Other sources arrive at an entirely different result. Quoting the UK-based Environmental Investigation Agency, Hoang et al. (2010) state that 50 % of the natural forest disappeared between 2000 and 2005.

Can we believe the official forest cover change data from Vietnam? Most observers praise Vietnam as a success story in having achieved a reversal of the typical path of a developing country where forest resources are depleted while the economy is growing (McElwee 2009; de Jong et al. 2006). Table 1 shows an increase in forest cover since 1995. Looking at the data concerning Quang Nam province compiled from four different sources and summarised in Table 2 it does leave the observer somewhat skeptical of the picture international organisations and the Vietnamese government want us to believe in. Namely, a story describing a country transitioning from a deforestation phase to a reforestation stage (UNDP 2011). From 1999 to 2008, the natural forest area remained constant in Quang Nam (third column).

Whether the observed data limitations in Quang Nam province singles out this province as an exception to the general picture of the quality of forest data in the country, or if all forest data in Vietnam are really figures based on uncertain estimates, demands a more genuine detailed study of changes over time in the entire country. This is a task which goes beyond the scope of this chapter. The story of a successful developing country changing from deforestation to afforestation could well be far from the reality.

It has been argued that the government-initiated reforestation programs 327[2] and 661[3] have played a central role in generating the reversal in the deforestation process (McElwee 2009). Forest restoration efforts have aimed at fulfilling the dual objectives of conservation and forestry-based socio-economic development (MARD 2001; Jong et al. 2006). Under program 661, the goal was to create two million hectares of protection and special use forest in addition to three million hectares of new production forest (Decision 661, 1998). Protection forest refers to an area which needs protection either in relation to water resources or the mitigation of soil erosion, while national parks and reserves are defined as special use forest. Finally, production forests are those available for pure exploitation of timber and non-timber products.

State Forest Enterprises, which were allocated more than 1/3 of the total forest area in 1998 totaling 4.4 million ha (de Jong 2006), underwent institutional changes, as their activities were seen as one of the central factors contributing to national deforestation. Due to rapid deforestation, the government imposed restrictions on logging quotas issued to the SFEs, beginning in 1990 (Sunderlin and

---

[2] Also known as the 'Greening the Barren Hills Program'. Henceforth it will be referred to as program 327.

[3] Also know as 'The Five Million Hectate Reforestation Program'. Henceforth it will be referred to as program 661.

Table 1 Change in forest cover in Vietnam, 1945–2002, in 1 m ha

Forest category	1943	1976	1990	1995	1999	2002	2004	2008
Total forest area	14.3	11.2	9.2	9.3	11	11.8	12.3	13.2
Plantation forest	0	0.09	0.7	1.1	1.5	2.0	2.2	2.8
Natural forest	14.3	11.1	8.4	8.3	9.5	9.9	10.1	10.4
Forest cover (%)	43.0	33.8	27.8	28.2	33.2	35.8	36.8	38.7

Sources de Jong et al. (2006); www.kiemlam.org.vn

Table 2 Forest area by classification of aggregated forest type in Quang Nam province, in 000 ha

Year	Total area	Forest area	Natural forest	Plantation forest	Non-forested	Barren
1999[a]	1,040	426	389	37	369	
2002[a]	1,040	445	388	57	350	
2004[a]	1,040	445	388	57	350	
1999[b]		426	389	37		
2002[b]		440	389	50		
2004[b]		449	389	59		
2007[b]		457	394	63		
2000[c]						
2007[c]	1,040	448	388	60	363	230
2008[d]	1,040	457	387	70	361	223

Sources
[a]Nguyen et al. (2009)
[b]Statistical Publishing House. Quang Nam Statistical Yearbook. (2008) Tam Ky
[c]Vietnam Government (2010)
[d]MARD (2011), www.kiemlam.org.vn

Huynh 2005). SFEs saw their allocated extraction quotas being reduced from 1.2 million m^3 in 1992 to 250,000 m^3 in 2003 onwards (Meyfroidt and Lambin 2009). Changing the status of the SFEs was seen as the solution for confronting the loss in forest cover. Whether the strategy was successful is debatable. Concerns are being raised that the reduction in SFE quotas either contributed to an increase in illegal national timber extraction (Sunderling and Huynh 2005) or illegal imports from neighboring countries (Meyfroidt and Lambin 2009).

The detrimental role of SFEs was acknowledged by the Vietnamese government through Government Decision 187 in 1999 which stated that only SFEs able to operate their own production and business activities would be given land use rights certificates as stated in the Land Law[4] (Jong et al. 2006; EASRD 2005, Government Decision 187). SFEs are now either operating as forest companies or as Protection Forest Management Boards (PFMBs). In Quảng Nam province, five different PFMBs are now active, and three forest companies are exploiting plantation forests and natural forests for production purposes. In order to achieve effective reform, the government allowed those SFEs, which carry out business and production activities, to operate under the market system while transforming SFEs carrying out public interest activities into PFMBs (EASRD 2005).

Critics have pointed out that Program 327, implemented in 1992 by a Council of Ministers' Decision, was heavily dependent on the proliferation of a few exotic tree species, such as Eucalyptus, Caribaea pine and Acacia (de Jong 2006). Eucalyptus is gradually being superseded by acacia species (e.g. Acacia auriculiformis, A. mangium) in domestic plantations due to the agronomic and economic benefits which accrue from these tree species. Acacia trees exhibit fast growth rates, are well-adapted to poor soil conditions and, as legumes, have nitrogen-fixation ability (Van Bueren 2004). Critics have highlighted several negative aspects of increasing areas under mono-culture species such as acacia, pine and eucalyptus. These include the lack of biodiversity value of these species and that mono-cultures might have adverse effects on the environment such as soil acidification and reduced stream flow (Clement and Amezaga 2007). In terms of increasing forest cover, the newest program, Program 661, seems also to have proven successful, and still observers question whether the achievements can be linked to the program, since many donor-backed programs are implemented simultaneously (de Jong 2006). Like its predecessor, Program 661 has also been criticised for its heavy reliance on exotic, monoculture species with an increasing focus on acacia (Barney 2005; McElwee 2009; Van Bueren 2004).

Case-studies indicate that degraded land which is not being efficiently used by SFEs can bring a certain income to households if the land is privatized (Nguyen et al. 2010). However, the option of investing in forest activities is constrained by access to capital, so poorer households might not always benefit from the forestry reform programs (Sikor and Nguyen 2007). A number of scholars have examined the importance of forest access to poor households, in times of crises. A study undertaken by Quang and Ahn (2006) demonstrates the importance of Non-Timber Forest Products (NTFP) for poor households, where the share taken up by NTFP in some cases is as important as agricultural income (30 % of the household income in one village). If that is the case, a sustainable reforestation policy should then allow collection of NTFP to be continued in forests in order to ensure poor households in remote areas a proper income. External shocks, like extreme weather conditions, cause households to engage in economic activities other than agriculture among which NTFP extraction has been ranked very high

[4] The process of SFE reform was further strengthened by the issuance of Decree 200 in 2004 (EASRD, 2005).

as a coping strategy (Völker and Waibel 2010). The authors conclude that forest policies which are too restrictive might limit the options of households to turn to an increased use of forest resources in times of crises (Ibid). Building on this observation, another outcome would be the deforestation or conversion of primary forest into plantation areas could engender similar negative effects and limit the options of shock absorption in times of crises.

2 Scope of the Chapter

This chapter analyses the impact of changes in land use and forest use on the household economy of different social strata (poor, middle income and rich households) comparing two communes, Tam Trà in Núi Thành district and Trà Tân in Bắc Trà My district. Both districts are located in Quảng Nam province in the South Central Coast region and field work was carried out over a two month period in 2011.

The analysis addresses five central research questions:

1. Is the area of plantations increasing, and if so is it at the expense of natural forest?
2. What is the household rationale for investing in forestry, compared to other land uses?
3. Who are the losers and the winners of the supposed plantation area expansion?
4. Has the acacia expansion been driven by government policies or market forces?
5. Is the rationale for changing forest use even valid in the context of an increasing risk of natural disasters?

2.1 Hypotheses

Our study has two central hypotheses. Firstly, we posit that poor households will not only benefit the least from an expansion of plantation areas, but a plantation economy could in fact also constrain their access to land and other resources in the long-term. Secondly, we suggest that household and government rationales for investing in particular land use systems is context-specific and depends on local conditions. We wish to illustrate this by selecting two study sites of varying sizes and different histories of government intervention. Finally, we wish to demonstrate how government intervention in land use and various households' rationales for investing has great significance in the context of climate change and natural disturbances.

Approach. As noted previously, poor households are most dependent on NTFPs as a livelihood component and a vital coping strategy. We therefore wish to investigate whether or not the natural forest cover in our two study sites has been converted to other land uses such as acacia or agriculture plantations. This is done by using satellite imagery analysis. We wish to compare changes in dense forest cover in the two communes through the interpretation of satellite images. The methods used for this classification are described in greater depth in Sect. 4.

We then investigate two of the land use types which have most likely replaced natural forest land. This is done through an assessment of the household rationale for pursuing agricultural and plantation strategies, respectively. We also assess the benefits and costs of both strategies in our study sites. A quantitative and qualitative analysis is employed for this purpose based on a household survey with closed- and open-ended questionnaires. We discuss the rentability of two land uses: wet rice farming and acacia forestry.

Due to the government's strong promotion of reforestation through the planting of mono-cultures we turn our attention to the actors who have chosen to engage in household forestry. In light of the status of local forest cover and the rationale of households for pursuing forestry, we wish to investigate *who benefits* and *who does not benefit* from this strategy and *why*. In our study area reliance on paddy rice farming has to a large extent been superseded by acacia production and acacia/cassava agro-forestry systems. In order to uncover the ways in which the introduction of acacia and the shift from agriculture to forestry has affected social/agrarian differentiation a number of qualitative field methods were employed in order to address the issue of winners and losers in the process of acacia production expansion. These methods include life history interviews, well-being rankings and village timelines. Our assessment of winners and losers in the plantation expansion then poses the question of whether the losers in this process have ended up as such due to the vagaries of market forces or as a result of the urging (coercion?) of State actors, and to what extent the modifications of the landscape are due to deliberate policy, pure household decision making or a combination of both.

Finally, we evaluate the soundness of both household and government rationales in the choice of land use given central Vietnam's status as a biophysically vulnerable and exposed region in Vietnam.

3 Central Concepts

The household and the various coping and livelihood strategies that its members employ are central to our analysis. A livelihood refers to the means of gaining a living which include capabilities and access to tangible and non-tangible assets (Chambers 1995). Tangible assets refer to stores of food and cash as well as livelihood resources—encompassing different forms of capital, e.g. natural, social, human, physical, and financial—while intangible assets refer to claims of support in times of stress and shocks as well access to livelihood resources (Scoones 1998; Chambers 1995; Chambers and Conway 1991). A livelihood is sustainable when it ensures the satisfaction of basic needs and the ability to secure against shocks and stresses (Chambers 1995). Where *stresses* are continuous, cumulative, and predictable pressures such as seasonal shortages and declining resources, *shocks* are impacts which are sudden, traumatic, and unpredictable such as natural disasters (Chambers and Conway 1991).

Household vulnerability constitutes a central aspect of our analysis. We here refer to social vulnerability as opposed to biophysical vulnerability (Brooks 2003; Adger et al. 2004). This choice is made for a number of reasons. Firstly, we seek to understand the underlying political and economic context in which certain households lose access to productive resources which enable them to cope with adversity. Secondly, we are assessing household vulnerability to multiple stressors and not simply those related to natural disasters although these may be dominant given our geographical focus. We have modified the definition of vulnerability formulated by Wisner et al. (2004)[5] to better suit our analysis. Our working definition of vulnerability is as follows:

> The socio-political context and the social and environmental assets which accrue to households through the processes associated with this context which influences households' ability to cope with and recover from changes in land use and natural disasters as well as the interplay between these two stressors. Households are vulnerable when they lack sufficient endowments of these assets and are exposed to stresses and shocks.

Following Scoones (1998) and DFID (1999) we identify two specific vulnerability contexts within which the command over endowments[6] and the ability to transform them into income and welfare, also referred to as entitlements (Leach et al. 1999; Wisner et al. 2004; Adger et al. 2004), is crucial to allowing either coping or adaptation to the stresses and shocks associated with the vulnerability context. Transforming structures (e.g. levels of government and organisations) and processes (institutions, laws, policies, and culture) have a profound influence on the vulnerability context and thus household vulnerability. In addition, the linkages between context, livelihood assets, and structures and processes influence existing livelihood strategies and can enable the creation of new portfolios of livelihood strategies (e.g. migration, diversification of income-generating activities, intensification of agriculture, off-farm labour) which produce livelihood outcomes (Scoones 1998; DFID 1999). Following from our definition of vulnerability two contexts are explored in this paper:

1. Land use change over a ten-year period with expansion of acacia as the main driver
2. Tropical storms with the potential to disrupt the production of acacia

4 Estimation of Forest Cover in the Two Communes

In order to estimate changes in forest cover in the past decade for the two communes Trà Tân and Tam Trà satellite remote sensing techniques were used. Using satellite remote sensing techniques and Earth Observation data for monitoring

[5] 'The characteristics of a person or group and their situation that influence their capacity to anticipate, cope with, resist and recover from the impact of a natural hazard'.

[6] By endowments we mean livelihood resources or capital forms, e.g. natural, social, human, physical and financial capital.

changes in land cover has been suggested as the most cost effective method that offers a spatially distributed data source consistent in time (Tøttrup 2002). Many different methods for detecting changes in land cover based on satellite remote sensing data have been suggested and tested for South East Asia (e.g. Yen, Ziegler, Huettmann and Onyeahialam 2005; Fuller 2006). In this case we performed a post-classification change detection (PCCD). In a PCCD data sets from different dates are compared in order to detect if changes have occurred in the periods between the acquisitions of each of the data sets. First each of the data sets is classified into meaningful land cover classes and subsequently the changes in the spatial distribution between the two images are analyzed.

Satellite images were first classified using a maximum likelihood classification based on the 6 spectral bands of the ETM scene in order to estimate the changes in dense forest cover in the two communes. Following this supervised classification, all areas classified as dense forest were isolated for further analysis.

Classification of dense, primary forest in South East Asia is notoriously difficult due to the many different land cover classes that potentially have a very similar spectral profile. Many different classification schemes have been suggested in the literature in order to address some of the problems encountered. Among these are various statistical tools and procedures that are applied either to the data before the actual classification, or are applied to the classification result (Tøttrup 2002, 2004). In this case a maximum likelihood classification based on 6 of the 7 original spectral bands was employed in order to classify land cover. Due to the limited spatial extent of the two communes in question the classification result was visually screened and controlled. Further, all areas smaller than 10 hectares in extent were removed from the results. Following the initial classification a moving 5×5 filter was applied in order to minimize single pixel effects and to smooth edges of the identified areas.

It is clear that in a classification of a 12 years old LANDSAT scene some errors of omission as well as commission can be expected. That is: Areas can be wrongly identified as dense forest (commission) or areas that should be identified as forest can be classified as another land cover type such as plantations (omissions). It is difficult—if at all possible—to find other reliable sources to use for ground truth, hence the analysis must be based on spectral signatures, statistics and detailed knowledge of the area today.

On the side of commissions, a closer evaluation of the classification result reveals that some commissions may occur in open forest areas affected also by shadow effects. We have attempted to alleviate the confusion caused by this phenomenon by a close manual scrutiny of all such areas, and manually omitting these from the dense forest class. Further, some densely forested areas were indicated in areas with a high cloud cover. These were also omitted in the final result due to the unreliability of the spectral signals due to high water content in the atmosphere. As can be seen in Table 3, the cloud cover in the two communes is not high, but many areas around the clouds were clearly affected by haze and high water content while at the same time being classified as dense forest. These areas were also removed from the further analysis, due to the unreliability of the classification result. Hence, corrections to the classification results have been performed based on close visual inspection. The corrections have all been performed on the basis of not overestimating the amount of dense forest in 2000, rather taking any possible error as an indication of unreliability and hence the removal from the following analysis.

Spatial extent of densely forested areas in 2010 was established using a visual classification of a Quick Bird WV-1 image (see also Okubo et al. 2010). Using both the pan chromatic and spectral bands of the QB image a very detailed outlining of all densely forested areas was performed. While visual classification methods have been accused of producing less reliable results (Lang and Langanke 2006) it has nevertheless been applied in a number of studies on land use and land cover changes. In order to comply with the classification of the LANDSAT scene, only larger, spatially coherent areas of more than 10 hectares were included in the analysis. Also, it should be noted that the very high spatial resolution of the QB image allows for a very detailed inspection, not just of spectral response functions of different surface types, but also for the inclusion of spatial arrangement, such as the often symmetrical layout of plantations in opposition to the densely forested areas. Considerations of spatial arrangement in the LANDSAT image is not possible to same degree and the classification of the LANDSAT image was performed based only on spectral signatures as described above.

Following the classifications of the two types of images, the spatial extent of densely forested areas in the two communes was calculated by a simple subtraction method.

As it can be seen in Table 3, the two communes are remarkably similar in their starting point with 12–17 % of the total area covered by dense forests. However, in the course of the decade the communes underwent very distinct developments in respect to amount of land covered in densely forest. In Trà Tân, the smaller and somewhat more remote commune approximately 15 % of the densely forested areas are found to be converted into other types of land cover. However, in Tam Trà almost half of the area covered in dense forest had been converted during the decade into other types of land cover than dense forest. Tam Trà is a special case

Table 3 Forest change in the two communes

	Hectares	
	Trà Tân	Tam Trà
Non-forest	1993	7070
Clouds	119	661
Forest 2010	453	1217
Forest removed	81	1137
	Percentage	
Non-forest	75.34	70.11
Clouds	4.51	6.55
Forest 2010	17.11	12.07
Forest loss	3.04	11.27
Percentage of forest loss	15.10	48.30

Source: LANDSAT images

since a large part of the natural area within its boundaries is what the government terms "protection forest". However, protection forest in the commune consists of both regenerating natural forest as well as forest planted with acacia.

Based on observations and interviews carried out during several field trips to both communes in 2011, it is highly evident that plantations are much more omnipresent in Tam Trà compared to Trà Tân. Semi-structured interviews with households in Tam Trà also confirmed that plantation increase in area and that this increase is exclusively Acacia plantations. We now turn to a comparison of wet rice and acacia systems in order to compare the profitability of each and the rationales which underlie the decision to engage in acacia production.

5 Profitability of Land Use Systems and Household Rationality

All farmers (65 households) expressed an interest in planting acacia trees in the two communes in which we conducted a household survey on land use changes. The size of the plantations varied from 1 to 10 ha, and for households who had already harvested the acacia trees, the gross revenue represented about 25 % of total income. In another study on farm-based plantation strategies in Vietnam, the scholars estimated the contribution to be on the order of 10 % (Sandewall et al. 2010). At this stage, we cannot explain the reason for the difference, but it could be a question of access to pulp mills, which are more accessible in Quảng Nam than in the three Northern provinces included in the study of Sandewall et al. (2010).

Only 2/3 of the interviewed households have harvested acacia trees in 2010, but all households, without exception, have invested in small-scale acacia plantations. Out of these households, three did not gain any income from agriculture. The economic rationale of the farmers to switch to plantation forestry makes sense, if only the input costs and cost of labor are included. Households with plantation holdings use a substantial amount of their own household labor force in pruning and weeding on the plantations, leading to a much lesser net income if calculated as costs (shadow pricing of household labor force). In economic terms, shadow pricing is useful to other types of occupation; however, in a social situation with few other alternative income sources, the calculation makes less sense.

On average, the gross income for households harvesting acacia in 2010 was 25 million Vietnam Dong (VND),[7] while the cost of household labour input was 10 million VND, with an additional cost of physical inputs[8] and hired labour of 2.3 million VND. Rice cropping yielded a gross income of 10 million VND with input costs (physical costs and labour sharing costs) of 0.7 million VND. Own labour costs averaged nearly 2 million VND.

[7] 1 USD = 20.000 VND.

[8] Chemical fertilizer, herbicides and fungicides.

Despite the fact that the rationality of the plantation investments could be questioned, using shadow pricing, poor households (20 million dong or less) have a firm interest in the plantation venture, since 15 % of their income comes from working on other households' acacia plots. On average, the poor households gain 3 million VND from acacia plantations and 1.8 million VND from working in the acacia plots of other families.

Another explanation for the interest in plantations could be that recently prices for wood products have increased rapidly. An even more important fact is the desire for households to obtain Red Book certificates legitimizing their land possessions. The clearing of land and subsequent planting of acacia trees in plantations make the former forest a de facto possession of the household. From the household survey and our in-depth interviews with household members, we cannot confirm that the applications for Red Book approval of forest areas (plantations) are accepted by the authorities. According to local government officials and staff not a single household in the entire commune, barring those who had participated in programs 327 and 661, had received a Red Book when we conducted our field work.

In the following, we will discuss the three main types of households in the two communes, their income base and the repercussions of the acacia plantation expansion on their livelihoods.

6 Livelihood Assets, Household Categories and Acacia Plantation Expansion

Using well-being rankings we distinguish between three household types: wealthier, middle and low income households. In the two communes we could identify common features for these three groups, but also differences, mostly likely linked to variability in land use and endowments of natural resources.

6.1 Wealthier Households

Wealthier households rely less on annual crop production than poorer households. Often their income comes from having employment which ensures a stable salary such as teachers or health workers. Income from these positions includes 'hardship allowances' for working in mountainous areas. While the wealthiest segment in all hamlets included those with stable salaries, they were not always those with the greatest endowments of natural capital. In Tam Trà this is partly due to the fact that teachers and local government staff do not have time to devote to the management of large forest areas. In Trà Tân, the wealthiest households are those with large areas of acacia, livestock, those with enough rice to feed their families and those that hire labour. The wealthiest households, in both communes, are able to hire labour to work on their acacia plots allowing them to spend their time on other activities. Furthermore, because the wealthier households took the lead in acacia expansion, they acquired the best land. Today, they have the financial resources to invest in the acacia expansion. These households primarily engage in forestry although this is more the case in Tam Trà than in Trà Tân where livelihoods in general are more diverse.

6.2 Middle-Income Households

Less wealthy groups devote 50 % of their time managing their own fields and forest plots and 50 % of their time working as hired labour. Households with a medium level of well-being rely on seasonal work such as working on coffee plantations in the Central Highlands. This work is carried out for 2–3 months a year. Medium households lack capital to expand and invest properly in their acacia production. Therefore many of these households enter into contracts with rich households. The rich households provide labour and capital and harvests are split 50/50 between the rich and the middle-income households.

6.3 Poor Households

Households in the least well-off group are often those with old, disabled, sick and weak household members. Furthermore, some rankings revealed that female-headed households as well widows and widowers frequently were in the lowest ranking groups. The least well-off households also lack rice to eat—many are landless and thus have to buy all their rice—and are unable to obtain loans even from the Social Policy Bank of Vietnam[9] due to their lack of collateral. The least well-off depend on casual labour including working on acacia plantations. Less well-off households are primarily engaged in agriculture and have experienced declining yields with bank erosion serving as an example of a natural disturbance.

7 Household Rationale for Engaging in Acacia Production: Government or Market?

Studies from other areas of Vietnam have investigated incentives and decision-making structures in the context of land use change. While a number of perceived benefits for local people to switch to forestry were present—government subsidies, collateral for loans, plantation products—the initial reception was not favourable (Dung and Webb 2008). This is due to the fact that the acacia species initially

[9] http://www.vbsp.org.vn/. The bank provides low-interest loans, credit and micro-credit schemes to poor households and groups in a range of sectors.

distributed were of low quality and that the land allocated was traditional swidden farming land (Dung and Webb 2008). However, this reluctance only lasted during the initial inception phase. Incentives started to increase as new high yielding plantation species such as Acacia mangium were introduced. These fast-growing species produced high yields, were more suitable for local conditions and hence more profitable for households (Dung and Webb 2008). The government strongly encouraged reforestation and forestland allocation was carried out.

In a similar study of land use decision-making in Vietnam, Clement and Amezaga (2008) show that while the majority of farmers invested in acacia very few invested in forest production as the preferred choice of land use. Rather, acacia production was the least-worst option in the wake of declining yields of annual crops and the subsequent breakdown of indigenous institutions for management of cropland (Clement and Amezaga 2008).

In Tam Trà land use changed significantly from 2000 onwards. Acacia was initially introduced through a government pilot project (the Agroforestry Extension Program) in 1995. Participating households were given seedlings and a salary for planting acacia. Under program 327 in 1998 and 661 in 2002, households were once again provided seedlings and technical training. Households were allowed to harvest after a fixed number of years (at that time 8 years) and under Program 327 the entire plot could be harvested all at one time. Under Program 661 harvesting regulations were changed and households could only harvest up to 50 % of their plots at a time. After an additional 2 year period, the remaining trees could be harvested. Under both Program 327 and 661 households could harvest according to regulations specified by the Phú Ninh PFMB. Several households stated that the introduction and expansion of acacia coincided with the implementation of these programs.

Acacia is now harvested on former Program 327 areas, Program 661 areas, although cutting of pine and takian (*Hopea odorata*) is not allowed, and private areas. Once the programs came to an end, households started encroaching on the natural forest to replace it with acacia trees.[10] While local people have cleared and claimed natural forest areas in order to plant acacia, a respondent declared that the introduction of the tree has improved the lives of people in the commune.

Another respondent in Tam Trà stated that the reason for investing in acacia was poor soil quality which means that yields of upland rice and cassava were too low. This is in line with the conclusions drawn by Clement and Amezaga (2008). The implementation of program 661 in Tam Trà also involved forest protection contracts. A man informed us that his household was provided with acacia seeds and money in exchange for protecting the forest they planted. However, since then, the general reaction from households has been to increase the area under acacia plantations, not to protect the forest. This has occurred under lax government enforcement of forest regulations. It appears that the government has encouraged households to plant acacia, but has cared little about enforcing the regulations (i.e. preventing transformation of the natural forest). However, some respondents noted a tougher stance on encroachment and stronger punishments since 2009 (see below).

[10] Both qualitative interviews and our household survey confirmed this trend.

While the government-supported programs which have been implemented since 1995 have resulted in an increase in the supply of acacia trees, several respondents confirmed the match between supply and demand. One respondent from Tam Trà claims that in 2001, he and other households started to invest in acacia themselves. That year a relative from a neighboring lowland commune told him to invest because there were several paper factories which would need the raw material from acacia. In his case, he did not know acacia would become such a valuable source of income program 327 was implemented in the late 1990s.

In Trà Tân, the situation has unfolded somewhat differently from that of Tam Trà. The vice-director in Trà Tân commune stated that acacia trees were introduced officially to the commune in 2005, and households claimed that investment in acacia was purely a response to price signals of the market, since the government had not implemented any program promoting acacia for either production or protection purposes. In one hamlet in the commune, acacia trees belong to both local inhabitants and outsiders. According to them acacia was introduced in 2005 by people from lowland areas. They had attended hamlet meetings where plans to plant rubber had been discussed but acacia was never been mentioned in any official meetings.

Another respondent noted that almost all households he knew of had planted acacia and that the tree production is highly profitable. However, the same respondent also claimed that investing in acacia is costly and the price given for the harvested trees is dependent on the proximity of the selling site to the main road. We see from the household survey that the mean income/costs ratio is twice as high in Trà Tân compared to Tam Trà (a mean of 1.9 respectively 0.7). In line with the respondent's claim, the variation is significant in Trà Tân (a standard deviation of 3.3 versus 0.8 in Tam Trà).

We conclude that the acacia plantation was a mix of government support programs and market forces, illustrated by the two differing cases. Enforcement of environmental rules is lacking.

8 Vulnerability in the Context of Climate Variability and Change

Compared to the wealthiest households in the communes, the middle-income and poor households still rely to a large extent on paddy rice cultivation. Paddy rice yields are declining and the main constraints to production are flooding and bank erosion. Bank erosion responses can be triggered both by natural hazards such as flooding and heavy rainfall events as well as by land use. People who have fields located near the riverbank are in some cases abandoning their land due to the poor, eroded and sandy soils which constrain rice production.

Household responses to low paddy yields can be observed in the form of two main coping strategies. Since rice production for many households is becoming increasingly insufficient to cover subsistence needs some households turn to casual work on acacia plantations. In addition, the declining rice yields cause households to clear plots in order to plant acacia—imitating wealthier households—in order to earn income. The pressure on the forest and lack of land has increased to such an extent that households have started to reclaim land in protection forest areas managed by the Phú Ninh PFMB.

One respondent cleared part of a protection forest area and planted acacia. The Phú Ninh PFMB subsequently fined him 4 million VND while forcing him to abandon his plot. According to him: 'In 2009 the restrictions on forest clearing became more serious. Before that time we could clear as much as we wanted'.

Low agricultural yields and a reliance on casual labor leave many households relatively vulnerable when impacts of tropical storms are experienced. A growing population of vulnerable households in need of food and income may exacerbate the clearing of natural forest in the long-term. The Ketsana typhoon had an immediate negative impact which mainly destroyed acacia plantations. Due to the effects of Ketsana most households were forced to harvest earlier than expected. For less well-endowed households this meant not being able to retrieve seeds to plant for the next period. One respondent reported that she had to buy the seeds instead. Ketsana strongly affected her household since 50 % of her trees were damaged. This constituted a significant constraint since she had taken a government loan to invest in acacia. People with bigger areas were affected more by the storm in terms of direct damages and losses. Results of wealth rankings indicate that the least wealthy groups in the hamlets—primarily made up of households who rely on work as hired labour—are the most vulnerable despite not having the greatest material losses. During shock events such as Ketsana poor households and the landless are arguably most at risk since they depend on short-term income. Since they rely mainly or exclusively on casual work on plantations, a halt in production can have highly detrimental impacts on their ability to secure income and buy food. In addition, these households also have little or no land to generate food for subsistence. Though the rich households lost the most in terms of physical assets, they were also those that recovered the fastest.

In a longer perspective the poor households are the most vulnerable, since in a time of crisis, they lose their income from casual work in the plantations, and they do not have the option to use the forest resources such as NTFPs, as they have potentially disappeared as a result of the expansion of acacia plantations. The option of turning to the natural forest in crisis periods is being curtailed, as the plantation area is substituting the natural forest. External shocks, like extreme weather conditions, cause households to engage in other economic activities, among which forest product extraction ranks very high (Völker and Waibel 2010). If the natural forest vanishes over time, this option is not viable in the future. In the two communes, the use of NTFPs was limited in the past. Only six households collected timber for house construction during the time of easy access to primary forests. Disappearance of primary forest could be less of a problem in the two communes, but in general households have short memories, and since no all forest is vanished, we cannot reject the hypothesis that the absence of access to NTFPs would not be a problem in the future.

9 Conclusion

Classification of digital images from Tam Trà and Trà Tân communes indicate that secondary forest areas have increased while primary forest has receded. Households in the communes expressed similar observations, claiming that the natural forests are being replaced by plantations. Household rationale for investing in acacia production differs in the two study locations. While Trà Tân has experienced limited, if any, government intervention in terms of reforestation, successive reforestation programs was implemented in Tam Trà which have focused on the dissemination of acacia species. This has had a profound impact on household adoption of acacia production. Conversely, the spread of acacia mono-cultures in Trà Tân commune was primarily a response to market forces.

Those who have benefitted most from acacia plantations are households with greater endowments of social and natural capital. Conversely, the poor and landless have been forced to take up casual work on wealthier households' plantations as a means to generate income to cover household needs. The social vulnerability of these groups cause them to cut down trees in protected forest areas in order to plant fast-growing acacia species in order to generate income to cover household needs. The PFMB has taken notice in recent years and these households are receiving heavy fines. Without alternative income sources, however, this survival strategy will most likely continue to be pursued especially since proper enforcement would require significant resources.

One dramatic natural disturbance event has revealed the sensitivity of the acacia land use system. The Ketsana storm caused massive tree collapse due to high winds and ensuing stem-break of the fast-growing acacia species. Since all households in one form or another are involved in this venture, the sustainability of livelihoods following this trajectory has been compromised.

References

Adger WN, Brooks N, Bentham G, Agnew M, Eriksen S (2004) New indicators of vulnerability and adaptive capacity. Tyndall Centre Technical Report 7 Tyndall Centre for Climate Change Research Norwich, UK

Barney K (2005) China and forest trade in the Asia-pacific region. Forest trends. Washington DC, USA

Chambers R (1995) Poverty and livelihoods: whose reality counts? Environ Urbanization 7(1):173–204. doi:10.1177/095624789500700106

Chambers R, Conway GR (1991) Sustainable rural livelihoods: practical concepts for the 21st century. IDS discussion paper 296. Institute of Development Studies. University of Sussex. Brighton

Brooks, N (2003) Vulnerability, risk and adaptation: a conceptual framework. Tyndall Centre Working Paper No. 38

van Bueren M (2004) Acacia hybrids in Vietnam: ACIAR project FST/1986/030. Impact assessment series report No. 27. Australian Centre for International Agricultural Research, Canberra

Clement F, Ameza ga JM (2007) Linking reforestation policies with land use change in northern Vietnam: why local factors matter. Geoforum 39(1):265–277
DFID (1999) DFID sustainable livelihoods guidance sheets. London, UK
Dung NT, Webb E (2008) Combining local ecological knowledge and quantitative forest surveys to select indicator species for forest condition monitoring in central Viet Nam. Ecol Indic 8:767–770
EASRD (2005) State forest enterprise reform in Vietnam. Technical Note. World Bank, Washington D.C
Fuller DO (2006) Tropical forest monitoring and remote sensing: A new era of transparency in forest governance? Singap J Trop Geogr 27 (1):15–29
Hoang MH, Do TH, van Noordwijk M, Pham TT, Palm M, To XP, Doan D, Nguyen TX, Hoang TVA (2010) An assessment of opportunities for reducing emissions from all land uses—Vietnam preparing for REDD. Final National Report. ASB Partnership for the Tropical Forest Margins, Nairobi
Hieu PS, Thuy VH, Thuan PD (2011) LeTV. Main Characteristics of statistical data and the statistical system for wood and wood-processing products in Vietnam Small-scale Forestry 10:185–198
Jong WD, Sam DD, Hung TV (2006). Forest rehabilitation in Vietnam: histories, realities and future. CIFOR, Bogor
Karsenty A (2006) Study on the impacts of market and investment liberalization on Vietnam's pulp and plantations sector, with particular reference to China. Vietnam Report. CIFOR, Bogor
Lang S, Langanke T (2006) Object-based mapping and object-relationship modeling for land use classes and habitats. Photogrammetrie, Fernerkundung, Geoinformation10 (1): 518
Leach M, Mearns R, Scoones I (1999) Environmental entitlements: Dynamics and institutions in community based natural resource management. World Dev 27:247–255
Ministry of Agriculture and Regional Development (MARD) (2011) www.kiemlam.org.vn
Ministry of Agriculture and Regional Development (MARD) 2010. Vietnam Forest Sector Progress Report 2006–010, Hanoiwww.vietnamforestry.org.vn
McElwee P (2009) Reforesting "bare hills" in Vietnam: Social and environmental consequences of the 5 million hectare reforestation program. Ambio 38(6):325–333
Meyfroidt P, Lambin EF (2009) Forest transition in Vietnam and displacement of deforestation abroad. PNAS 106(3):16139–16144
Ministry of Agriculture and Rural Development, International Cooperation Department, 5MHRP Partnership Secretariat (2001) Five million hectare reforestation program partnership synthesis report. Hanoi
Nguyen BN, Nguyen QT, Sunderlin WD, Yasmi Y (2009) Forestry and poverty data in Vietnam. In: The rights and resources initiative (Washington). The Regional Community Forestry Training Center in Bangkok and Vietnam Forestry University, Hanoi
Nguyen TT, Bauer S, Uibrig H (2010) Land privatization and afforestation incentive of rural farms in the northern uplands of Vietnam. For Policy Econ 12:518–526
Okubo S, Parikesit Muhamad D, Harashina K, Takeuchi K, Umezaki M (2010) Land use/cover classification of a complex agricultural landscape using single-dated very high spatial resolution satellite-sensed imagery. Can J Remote Sens 36(6):722–736
Poffenberger M (ed) (1998) Stewards of Vietnam's upland forests: A collaborative study by the Asia forest network and the forest inventory and planning institute. University of California, Center for Southeast Asia Studies, Berkeley
Quang DV, Anh TN (2006) Commercial collection of NTFPs and households living in or near the forests: case study in Que, Con Cuong and Ma, Tuong Duong, Nge An, Vietnam. Ecol Econ 60:65–74
Sandewall M, Ohlsson B, Sandewall SK, Viet LS (2010) The expansion of farm-based plantation forestry in Vietnam. Ambio 39:567–579
Scoones I (1998) Sustainable rural livelihoods: A framework for analysis. IDS Working Paper 72. Sussex, UK

Sikor T Nguyen TQ (2007) Why may forest devolution not benefit the rural poor? Forest entitlements in Vietnam's central highlands. World Dev 35:2010–2025

Statistical Publishing House. Quảng Nam Statistical Yearbook. 2008. Tam Ky

SunderlinWD Huynh TB (2005) Poverty alleviation and forests in Vietnam. Bogor, CIFOR

Tøttrup C (2002) Deforestation in the upper Ca river basin in north central Vietnam: a remote sensing and GIS perspective. Geographica Hafniensa, Copenhagen

Tøttrup C (2004) Improving tropical forest mapping using multi-date Landsat TM data and preclassification image smoothing. Int J Remote Sens 25(4):717–730

Vietnam Governement (1998) Decision 661, by the Prime Minister 1998. On objectives, tasks, policies and organization for the establishment of five million hectares of new forest. No. 661. Hanoi

Vietnam Government (2010) Department of forestry forest inventory, 2010. Tam Ky

Völker M, Waibel H (2010) Do rural households extract more forest products in times of crisis? evidence from the mountainous uplands of Vietnam. For Policy Econ 12(6):407–414

Wisner B, Blaikie P, Cannon T, Davis I (2004) At risk: natural hazards, people's vulnerability and disasters (2nd edn). Routledge, London

Yen P, Ziegler S, Huettmann F, Onyeahialam AI (2005) Change detection of forest and habitat resources from 1973 to 2001 in bach ma national park, Vietnam, using remote sensing imagery. Int Fore Rev 7(1):1–8

Climate Change's Impact on Natural Hazards in Quang Nam Province, Mid-Central Vietnam

An Duc Le and Lan Thi Thu Vu

Abstract This chapter discusses the natural hazards of flooding and drought and provides a short reference to landslides. Flooding might present a greater threat to the livelihoods of local people than drought. As yet we cannot fully understand the interplay between these natural hazards and other environmental factors, and it is difficult to establish a positive relationship between climate change and hazards. However, if the relationships are assumed to exist, projections are possible to draw up for future changes in flooding and drought. The chapter describes projections based on expected changes in rainfall, which are in accordance with the climate change scenarios developed for Quang Nam and Mid-central Vietnam by the Ministry of Natural Resources and Environment. Under a climate scenario in which there is an increase of 2 °C in temperatures by the end of the century, the flooded area in the province might increase by 20 %. Though mitigation plans are drawn up, no long-term strategy is yet in place to deal with an increase in the flooded area of this magnitude.

Keywords Flooding • Droughts • Costs of natural disasters • Climate change scenarios • Predictions of future changes in flooding and droughts

1 Introduction

Quang Nam province covers an area of 10,406 km^2 and has a population of 1.4 million inhabitants (Population Census in 1 April 2009). The province is located on the coastal region of Mid-Central Vietnam (Fig. 1), and it is considered a key economic development zone of the Central region. Quang Nam province is the home of two of the world's heritage sites: the My Son temple complex and Hoi An old town, as well as the globally recognized biosphere reserve Cu Lao Cham.

A. Duc Le (✉) · L. Thi Thu Vu
Institute of Geography, Vietnam Academy of Science and Technology, Hanoi, Vietnam
e-mail: leducan@yahoo.com.vn

O. Bruun and T. Casse (eds.), *On the Frontiers of Climate and Environmental Change*, Environmental Science and Engineering, DOI: 10.1007/978-3-642-35804-3_5,

Fig. 1 Administrative map of Quang Nam

Despite the many advantages of the province in terms of natural and human resources for socio-economic development, its economic growth rate is still low compared to other provinces in Vietnam. Furthermore, the province is challenged by a range of different natural disasters. In recent years, possibly due to the impacts of global climate change and socio-economic development, natural disasters have increased in Quang Nam causing more serious damage. Dry spells, flooding and landslides are currently the most serious natural hazards causing damage to socio-economic activities and the environment in the province.

This chapter provides a brief description of the status of these three natural hazards in Quang Nam during the past decade. Further, the chapter discusses possible relationships between climate change and the selected natural disasters. Finally, this chapter will attempt to produce general predictions concerning two of the three natural disasters (floods and droughts) by the end of this century within the context of global climate change.

Table 1 shows the damages and losses caused by natural disasters. Though the number of fatalities and missing persons are high (517 persons in a decade), Vietnam has been spared of major collapses in the country's economic and social development model. Vietnam is high on the list of disaster ridden countries, but ranks much lower on the list of fatalities (see "Impediments to Climate-Induced Disaster Management: Evidence from Quang Nam Province, Central Vietnam").

Table 1 Costs of natural disasters from 1998 to 2007

Years	Deaths/missing people	Injured people	Land/rockslide volume (m^3)	Property damage (billion in VND)
1998	54/1	36	1,040,000	390.0
1999	118/0	339	1,500.000	758.0
2000	13/0	0	450,000	139.3
2001	13/1	9	180,000	75.7
2002	0/0	0	Not significant	2.2
2003	32/2	5	150,000	91.4
2004	19/23	13	680,000	155.9
2005	12/5	24	440,000	109.7
2006	176/1	562	470,000	1,900.6
2007	47/0	339	$\gg$300,000	2,000.0
Total	484/33	1,327	5,210,000	5,622.8

Source Quang Nam People's Committee (2010) Plan for implementing project to improve community awareness and natural disaster management based on community in Quang Nam up to 2020.

Photo 1 Flooding in Hoi An city, 2009

During the observed decade, economic damages have increased, whereas the magnitude of landslides, regardless of origin, has followed a more erratic path.

Measured in terms of landslide volume, 1999 stands out as the worst year. Much of landslide damages occurred in the mountainous districts and flooding occurred in lowland districts, especially some areas such as Nui Thanh, Dai Loc, Dien Ban, DuyXuyen districts and Hoi An and Tam Ky cities. These were flooded to a depth of 1–2 m for many days (Photo 1). The landslide blocked almost all the roads in Quang Nam that year. Highway 1A and the North–South railway were flooded deeply under water for 1–2 m. In 2007, 125–233 communes in Quang Nam consisting of approximately 200,000 households were flooded, the communication and electricity systems were interrupted in many places, the roads were obstructed, highway 1A was closed for 40 h, more than 1,500 houses were destroyed and swept away, much infrastructure and facilities were damaged and hundreds of thousands people needed to be provided with food and other necessities. The problem of landslides is caused by several factors, including climate change. Isolating the impact of climate change is virtually impossible in an analysis of the root causes of landslides, and as a consequence the following discussion concentrates on suggested links between climate changes and flooding or dry spells

2 Drought or Dry Spells, Present Situation

Dry spells cause serious damage to agriculture in Quang Nam. In 1998 the drought lasted from May to September, affecting 32,767 ha of cultivated land, of which 2,819 ha was totally lost resulting in a lack of clean water for 200,000 people. In 2001, there was a /there were droughts in July and August, causing a decrease of 60 % of rice productivity in 8,000 ha and lack of water for many sectors, especially industry due to deep saltwater intrusion into the plains. For the years from 2002 to 2010, 10–15 % of the cultivated area was affected by drought. Drought also contributed to forest fires which increased the number of sandstorms covering fields and villages, which in turn led to soil degradation.

On average, the rainfall in the low flow period (from January to September) accounts for only 25–30 % of the annual rainfall and the long period without rainfall is considered a factor in causing drought on a large scale. For Quang Nam, its coastal plains suffer, on average, 9–10 long no-rain periods each year, at most 14 periods and at least 3 periods and each period lasts for about 14–17 days, the longest period ever seen is 100 days, that is to say, three consecutive months of hot and sunny weather without any rainfall. Its midlands and highlands experience about 5–6 long no-rain periods every year, at most 9 periods and each period lasts for about 9–10 days on average and the longest ever seen is 55 days.

Vietnam's Institute of Meteorology, Hydrology and Environment developed a drought classification system, classifying drought into 8 levels (Table 2), using the length of drought season to determine the drought level.

Table 2 Drought classification

Drought level	Description	Number of drought months
0	No drought	0
1	Very slight	1
2	Slight	2
3	Moderate	3
4	Strong	4
5	Very strong	5
6	Serious	6
7	Extremely serious	7, 8, 9

Quang Nam province experiences three dry months every year. In years of normal weather conditions, the dry period begins in February and April and ends with the Xiao-man (minimum periodic) rain which arrives in May and June. During the years with dry spells, the dry period cover the months from January to July, or even August.

During the last 50 years, dry spells, defined as months with less than 100 mm of rainfall, increased by 0.219 month (6–7 days) per decade. Most severe droughts during the summer-autumn season were seen in the years 1952, 1969, 1983, 1993, 1998, 2001, 2009 and 2010; during the winter-spring season in the years 1970, 1984, 2010 and the 10 years with slight or no drought: 1961, 1971, 1976, 1978, 1982, 1985, 1989, 2000, 2001, 2007.

3 Flooding, Present Situation

The topographical surface of Quang Nam is dissected by its dense river systems, including the Vu Gia-Thu Bon river system as the main system and the Tam Ky river system as a small river system. These two river systems are connected by the Truong Giang river.

Every year, Quang Nam receives 30.2 billion m3 of rain water out of which 21.5 billion m3 pours into the rivers and streams. Due to the steep terrain and the fan shape of the river networks in combination with high rainfall frequency and intensity floods formed in the rivers and streams in the Vu Gia-Thu Bon river basin exhibits typical features of mountainous floods: high intensity over a short period (both quick uprising and retreat), and large amplitudes and a sharp peak.

Flash floods occur annually in the upstream area and are a frequent threat to local people. Floods occur roughly 4–5 times a year, even 7–8 times in maximum in the Vu Gia-Thu bon river system. The highest floods are seen in October and November. They are typically caused by various weather patterns such as hurricanes, tropical depression, cold air and northeast monsoon resulting in heavy rains for many days when the soil became saturated with water from early floods and rains and water levels in rivers and streams were high already. In this basin, heavy flooding occurred at high frequency.

Floods in the Vu Gia-Thu Bon river system typically rises and retreats quickly and is of between 5 and 14 m and high (about 20–50 cm/h on average and 100–140 cm/h at maximum) in the upper and middle parts of the river. Further downstream withdraws happen less quickly, especially in tidal periods of tide. Consequently, most of flood run-offs overflow into fields in Ai Nghia and GiaoThuy to cause inundation over the whole downstream areas that cover the Dai Loc, Dien Ban, DuyXuyen districts and Hoi An city. In the years with heavy floods, the flooded downstream area fluctuated from 452.28 km^2 (1996) to 734.26 km^2 in 2007(see Fig. 2).

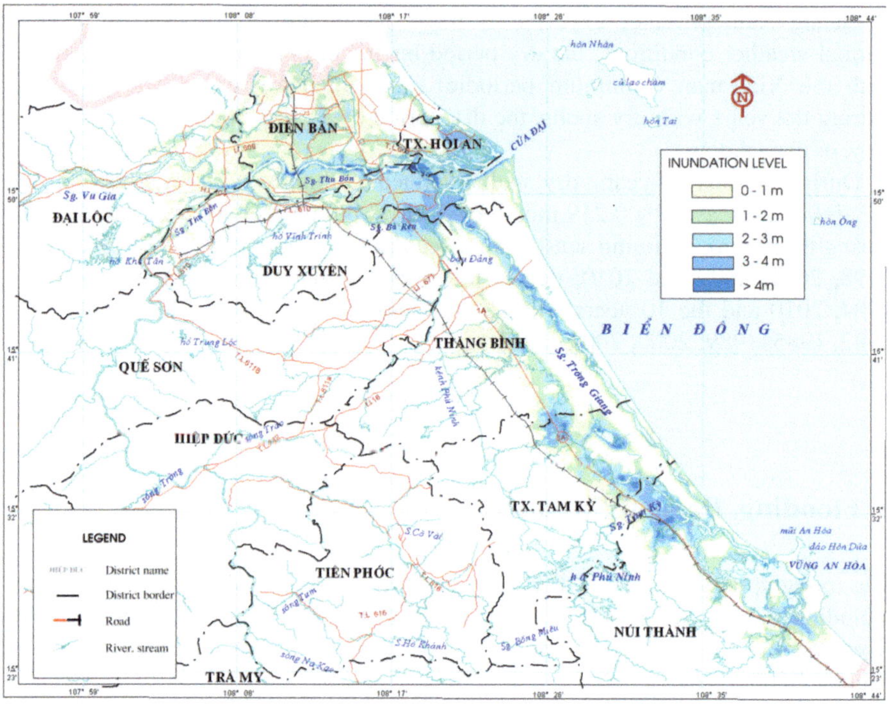

Fig. 2 Floods in Quang Nam (October 2007)

4 Climate Change Scenarios

According to the climate change scenarios for Vietnam (Ministry of Science and Technology 2009), the annual mean temperature of Quang Nam might reach an annual increase of 0.3 °C by 2020, 0.7–0.8 °C by the middle of the 21st century and 1.5–1.9 °C by the end of the 21st century under the medium and high scenarios. Winter temperatures could increase faster than in summer and autumn (September to November, cf. Table 3).

In order to describe flooding predictions in Quang Nam, projections are based on expected changes in rainfall, which are in accordance with the climate change scenarios developed for Quang Nam and Mid-central Vietnam by Vietnam's Ministry of Natural Resources and Environment (2009) and the calculations in one of the Denmark-Vietnam cooperation sub-projects (Van Tan 2010). We assume that changes in other natural factors and human impacts remain insignificant from 2012 to the end of this century. Though these assumptions are not realistic, the predictions show the pure impact on flooding of climate changes, and can be used as reference scenarios.

The reliability of the outputs calculated by the model obviously depends on the accuracy of terrain model. The higher the resolution of topographical data, the more accurate the results will become. The relief maps of the lowland at scale

Table 3 Increase of mean temperature of Quang Nam (°C) compared to 1990 under emission scenarios

Scenario	Period	2020	2050	2100
Medium emission	Whole year	0.3	0.8	1.5
	Winter (from December to February)	0.3	0.8	1.5
	Spring (from March to May)	0.3	0.8	1.6
	Summer (from June to August)	0.2	0.6	1.2
	Autumn (from September to November)	0.3	0.9	1.7
High emission	Whole year	0.3	0.7	1.9
	Winter (from December to February)	0.3	0.7	1.8
	Spring (from March to May)	0.3	0.8	2.0
	Summer (from June to August)	0.2	0.6	1.5
	Autumn (from September to November)	0.4	0.8	2.1

Table 4 Flooded area in km^2 corresponding to the high emission scenario

		2020		2050		2100	
Flooding level	2007 (km^2)	Area (km^2)	Compared to 2007 (%)	Area (km^2)	Compared to 2007 (%)	Area (km^2)	Compared to 2007 (%)
1. 0–1 m	177.5	173.2	−2.40	175.74	−0.99	177,59	0.05
2. 1–2 m	191.8	191.9	0.05	205.13	6.95	209.79	9.38
3. 2–3 m	178.0	178.2	0.11	195.46	9.81	195.07	9.59
4. 3–4 m	129.1	140.2	8.60	155.78	20.67	201.12	55.79
5. >4 m	57.86	60.32	4.25	72.18	24.75	102.75	77.59
Total	734.26	743.82	1.30	804.29	9.53	886.32	20.71

1/25,000 and of mountainous area at scale 1/50,000 have been used to develop the terrain model with cell size of 30 m.

The flooded area predicted for the years 2020, 2050 and 2100 compared to that in 2007 is shown in the Table 4. The figures show that the flooded area tends to increase by 1 % by 2020–20 % by 2100. The change in the flooded area is, however, very different between the various flooding levels. The flooded area at level 1 (flooded to a depth of under 1 m) will not change much or even decrease, but it will increase rapidly at level 5, the flooded area of 100-year flood in 2100 will increase by 77 % compared to the 2007 level.

The above-mentioned drought trend in Quang Nam leads to the prediction that by 2020 the length of drought season will increase 12 days compared with that in 1990 under the high emission scenario and 7 days under the medium emission scenario (Table 5); by 2050 it will increase by 25 days under the high emission scenario and

Table 5 Predicted increase in length of drought season (day) due to climate change compared to 1990

	Temperature increase			Change in rainfall			Total		
Scenario	2020	2050	2100	2020	2050	2100	2020	2050	2100
Medium emission	7	17	34	0	−1	−2	7	16	32
High emission	12	26	67	0	−1	−3	12	25	64

16 days under the medium emission scenario. In year 2100, the increase will be 64 days under the high emission scenario and 32 days under the medium emission scenario.

According to the drought classification, Quang Nam will be affected by drought at moderate level (according to both scenarios) by 2020; at moderate-to-severe level by 2050 and at severe-to-extreme level by 2100 under medium-to-high emission scenarios. Drought here will often occur in spring and summer with an average length of about 2.5–3.5 months.

5 Conclusion

Quang Nam is annually affected by different natural hazards, including flooding, dry spells and landslides causing great damage to socio-economic activities and the environment. In the 10 years from 1998 to 2007, these natural hazards caused 500 deaths and removed 5.2 billion m3 of soil and rock along the roads and cost 5,600 billion in VND of property damage. Following government guidelines "Active prevention and adaptation for development", Quang Nam has implemented a number of mitigation and adaptation measures such as raising community awareness of natural disaster prevention, improving management of the river basins, building facilities and infrastructures to withstand natural disasters, developing plans for evacuation of people from the frequently affected areas. Although the preparedness for natural hazards has been strengthened, there is no long-term strategy to cope with the scale of the natural hazards in coming decades, in the event that our predictions for increase in flooding areas and drought problems should prove correct. Solutions to prepare for and adapt to the natural hazards under impacts of climate change, include improvements in forecasting and warning capacity, management, and raising awareness among local people. Nevertheless, an increase of 20 % in flooded areas, under the high emission scenario, might call for more drastic adaptation measures.

Existing studies on relationship between climate change and natural hazards are still simple and lack sound evidence of the presupposed relationship. Our study is a preliminary one. Thus, a real need is well founded to improve the research and discussions on possible relationship between climate change and natural hazards, based on historical statistical data on rainfall and temperature and refined modeling of future scenarios.

References

Ministry of Natural Resources and Environment (2009) Climate change and sea level rise scenarios in Vietnam. Hanoi

Quang Nam People's Committee (2010) Plan for implementing project to improve community awareness and natural disaster management based on community in Quang Nam up to 2020. Tam Ky

Van Tan P (2010) Study on developing detailed climate change scenarios for mid-central Vietnam. Subproject 3, project assessing impacts of climate change on physical, environmental conditions and socio-economic development in mid-central Vietnam, Vietnamese-Danish research project, Hanoi

Impediments to Climate-Induced Disaster Management: Evidence from Quang Nam Province, Central Vietnam

Olivier Rubin

Abstract This chapter aims to identify deeper rooted socio-political impediments for effective natural disaster responses in Vietnam by drawing on first-hand evidence from the province of Quang Nam. Vietnam suffers from relatively high incidents of natural disasters due to the country's high susceptibility to natural hazards, and Quang Nam is found to be particularly vulnerable. There are, however, several socio-political obstacles to effective disaster management in the province. While decentralization processes have accelerated in Vietnam during the last decades, the limited devolution of power to the local political entities make them first and foremost instruments for policy implementation. The relationship between communes and local government institutions, therefore, is not symbiotic but rather driven by the central government. Concretely, disaster management meetings in Quang Nam were used to disseminate instructions and information from above. Vulnerable households, with an underdeveloped social asset-base, did not perceive disaster protection as a right, and they did not consider disaster protection as a legitimate demand on the state. The chapter argues that effective disaster management needs to extend beyond the policy-level to consider more fundamental shifts in the political structures and institutions.

Keywords Disaster management • Vulnerability • Fatalities • Costs of natural disasters • Disaster compensation • Technocratic bias

1 Introduction

Vietnam suffers from relatively high incidents of natural disasters due to the country's high susceptibility to natural hazards. Ninety-five percent of these natural disasters are climate-related (triggered by meteorological and climatological factors). Hence,

O. Rubin (✉)
Department of Environmental, Social and Spatial Change,
Roskilde University, Roskilde, Denmark
e-mail: olivierrubin@gmail.com

O. Bruun and T. Casse (eds.), *On the Frontiers of Climate and Environmental Change*, Environmental Science and Engineering, DOI: 10.1007/978-3-642-35804-3_6,

increased future susceptibility to climate change might have a considerable impact of Vietnam's disaster vulnerability. In that light, there is a need to focus on the socio-political obstacles to effective disaster management in the long term. Quang Nam is identified in this chapter as being particularly vulnerable to natural disasters, and the province therefore constitutes a purposely selected case, from which key qualitative findings will form the basis of broader socio-political analyses.

The socio-political analysis links findings on the ground in Quang Nam to meso and macro level political theory. The findings are based on interviews and surveys involving vulnerable households and local political officials in Quang Nam, and they are used inductively as a catalyst and guide for the subsequent political analysis. The analysis places the most prominent qualitative findings into a larger socio-political context drawing on political and institutional theory.

Five deeper-rooted impediments to effective disaster protection are identified and discussed in this chapter; they are:

(i) the lack of a social contract between the Vietnamese citizens and the government in disaster protection;
(ii) the lack of electoral competition both nationally but also locally;
(iii) the lack of devolution of power in local disaster management;
(iv) the weak social asset-base at the local level; and
(v) the technocratic bias in disaster management and climate change research.

These impediments are interrelated and rooted in Vietnam's specific socio-political institutions and dynamics. The chapter concludes that while equitable economic growth coupled with various state-led investments in physical infrastructure will continue to play a vital role in bolstering resilience to climate-induced disasters, a more fundamental shift in the disaster management paradigm in the light of the future implications of climate change would demand a fundamental shift in the political structures and incentives.

2 Natural Disaster Vulnerability in Vietnam

An investigation into climate-induced disaster vulnerability in Vietnam first and foremost entails defining and operationalizing natural disaster vulnerability. A natural disaster can be defined as a temporary event, triggered by a natural hazard that overwhelms local response capacity and seriously affects social and economic development (Hodell et al. 1995). Thus, a natural hazard is a necessary but not a sufficient condition for a natural disaster to materialize, as the extent to which natural hazards turn into disasters depends crucially on the socio-economic context. Thus, natural disaster vulnerability is linked both to the physical exposure to natural hazards as well as to the socio-economic assets that determine impacts of the physical exposure on society as a whole or various vulnerable groups. Vulnerability to natural disasters can be understood as the degree to which the country (or region) is susceptible to, or unable to cope with, adverse effects of natural hazards (Wisner et al. 2004). Socio-economic development and political interventions can help societies coping with the adverse consequences of natural hazards.

Data on natural disasters in Vietnam is drawn from the Emergency Events Database, a cross-national database on natural disasters compiled by the WHO Collaborating Centre for Research on the Epidemiology of Disasters (EM-DAT 2012).[1] The data has been averaged over the period 2000–2010 to smooth out year-to-year fluctuations, and to provide an accurate picture of contemporary vulnerability. Out of the 181 countries included in the EM-DAT database, Vietnam is near the top of the list (7th place) with respect to natural disaster incidents; the country experienced a reported 89 natural disasters in the period. However, Vietnam's natural disaster mortality of 0.41 deaths per 10,000 places it much further down the list at 76th place. What is particularly striking is that nearly all of Vietnam's natural disasters are climate-related (triggered by meteorological and climatological factors); floods account for more than half the natural disaster mortality in Vietnam closely followed by storms accounting for nearly 40 % of the mortality rates (Fig. 1).

The strong impact of flooding is evident, as it accounts for close to 75 % of all the households affected by a natural disaster with typhoons a distant second. Floods are often slow-onset natural disasters that have a greater propensity for eroding people's livelihood than their lives. With regards to economic damages, storms account for 58 % of the damages while floods account for 37 %. Hence, floods are more lethal and affect more people but storms appear to cause

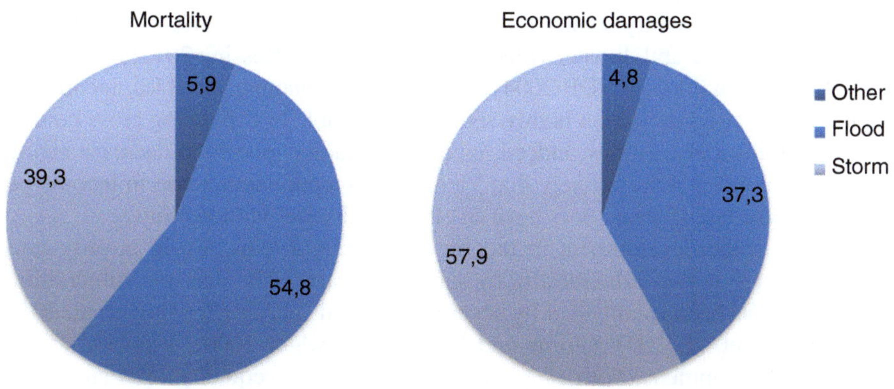

Fig. 1 Percentage of Vietnam's fatalities and economic costs based on natural disaster type 2000–2010. *Source* EM-DAT (2012)

[1] For a disaster to be entered into the database at least one of the following criteria must be fulfilled: (a) 10 or more people reported killed; (b) at least 100 people should be reported affected; (c) The government has declared a state of emergency; or (d) the government has called for international assistance.

more physical destruction. With such high share of climate-induced disasters (as opposed to geophysical disasters), Vietnam risks increased vulnerability in the future due to climate change.

Although Vietnam is rarely singled out as being particularly vulnerable to climate change in the IPCC 2007 reports, other publications—most notably the World Bank and the Asian Development Bank—do expect that Vietnam will suffer considerable adverse consequences of climate change. The World Bank study on the impact of sea-level rise on developing countries consistently ranks Vietnam as the first or second most vulnerable country across various dimensions (including land area impacted, population affected, and economic loss) to sea-level rise (Dasgupta et al. 2007). The Asian Development Bank reports Vietnam's average temperature rose by 0.14 °C per decade during 1951–2000 in particular due to hotter summers in recent years with average monthly temperatures increasing up to 0.3 °C per decade (Asian Development Bank 2009, p. 23). Vietnam's long coastline and multitude of river basins certainly could make it vulnerability to climate-induced hazards such as storms and floods (Cuong et al. 2009; Quy 2011). The impacts of climate change will be further compounded by Vietnam's environmental degradations; the destruction of mangrove forest, for instance, has left many areas more exposed to storm surges (Viner and Bouwer 2006). Hydro-power plants, in particular in central Vietnam, have been accused of causing or exacerbating flooding in recent years (Bruun 2012).

The upside to these dire projections is that natural hazards and natural disasters are only partially overlapping, so that an increased susceptibility to natural hazards will not necessarily translate into more natural disasters. Socio-economic characteristics appear to have a substantial effect on vulnerability to natural disasters. Many quantitative cross-country studies find that key socio-economic variables have significant impacts on natural disaster vulnerability (Rubin and Rossing 2012; Roberts and Park 2007; Kahn 2005; UNDP 2004). Of the socio-economic variables, income per capita has consistently been a highly significant variable in explaining cross-country variation in disaster fatalities. Indeed, income per capita appears to decrease natural disaster fatalities making it likely that Vietnam's exceptional increase in income per capita over the last 15 years does have an effect on disaster vulnerability.

Related to the development in income per capita, a reduction in poverty-rates also appears to lower vulnerability to natural disasters. The disaster vulnerability of the poor has been established by several empirical studies (Rossing et. al. 2010; Satterthwaite et. al. 2007 Simms and Reid 2006; Adger 2006). Vietnam‘s fairly equitable development (with only minor increases in inequality) has succeeded in reducing poverty rates (percentage living below $1.25 PPP/day) from 64 % in 1993 to 13 % in 2008 (WDI 2012). In a working paper, I have recently analyzed variations in Vietnam's disaster fatalities across provinces, and one of the key findings was that socio-economic variables accounted for between 50 and 70 % of the variation in disaster vulnerability even without accounting for the exposure to natural hazards (Rubin 2012).

In terms of institutional responses to natural disasters, the Central Committee for Flood and Storm control has been operational since 1955. The Ordinance on Flood and Storm Control in 1993 stipulates that at all levels throughout the country from central, provincial to district and commune, a Steering Committee for Flood and Storm Control shall be established. In emergency cases, the Prime Minister and the chairmen of the People's Committees of all levels may mobilize all forces, supplies and means of any organizations and individuals to rescue people, projects and properties threatened or damaged by floods and storms and shall take responsibility for their decisions.

More specifically with regards climate change, Vietnam ratified the United Nations Framework Convention on Climate Change (UNFCCC) in 1994 and the Kyoto protocol in 2002. Vietnam has compiled a National Strategy For Natural Disaster Prevention, Response and Mitigation to 2020 (Government of Vietnam 2007) with a heavy focus on adaptation efforts. Currently, the National Target Programme in Response to Climate Change drafted by the Ministry of National Resources and Environment (Government of Vietnam 2008) is being adopted, with several ministries such as Ministry of Agriculture and Rural Development, Ministry of Science and Technology, Ministry of Industry and Trade, Ministry of Transport outlining their action plans for climate change. In short, the Vietnamese government has given preference and gradually increased budget for natural disaster prevention, response and mitigation (East Meets West 2009).

3 Research Design, Case-Selection and Information Gathering in Quang Nam

The socio-political analysis aims to links findings on the ground in Quang Nam to meso- and macro-level political theory. First, however, the relevance of drawing on Quang Nam in the research will briefly be addressed.

As mentioned above, there is substantial variation in vulnerability within Vietnam, and certain regions are more vulnerable to natural disasters than others. The empirical evidence suggests that Quang Nam is particularly vulnerable to natural disasters. The Disaster Management Information System (DMIS 2011) has disaggregated natural disaster mortality down to provincial level.[2] The map below, based on DMIS data, illustrates the spatial distribution of natural disaster fatalities per 10,000 in the period 2000–2009. What is particularly striking is the apparent lack of a clear pattern in the spatial distribution of disaster vulnerability. Hotspots of natural disaster vulnerability can be found in Southern, Central and Northern Vietnam; they can be found in coastal provinces as well as mountainous in-land provinces; and they can be found in provinces with great rivers basins. The six most vulnerable provinces that are highlighted in dark red (Quàng Ngai, ên Bái,

[2] The database is compiled and maintained by the Central Committee for Flood and Storm Control (CCFSC) under the Ministry of Agriculture and Rural Development (MARD). The data is based on a template with a multitude of indicators for disaster consequences. At each level, province, district, commune and village there is a committee on flood and storm control that gathers and collates data before sending it to the level above.

Quảng Nam, Lao Cai, Phú Yên and Dồng Tháp) account for 37 % of all the natural disaster fatalities in Vietnam (in the period 2000–2009), although they only account for 8 % of the population. The population in these six provinces is more than seven times as vulnerable to natural disasters as the rest of Vietnam (2.9 deaths per 10,000 compared to 0.4) (Fig. 2).

When it comes to natural disaster damages per capita, other provinces report higher losses than do the six provinces. In particular, the city-provinces of Da Nang and Ha Noi report high losses (8 and 9 million Dong per capita,

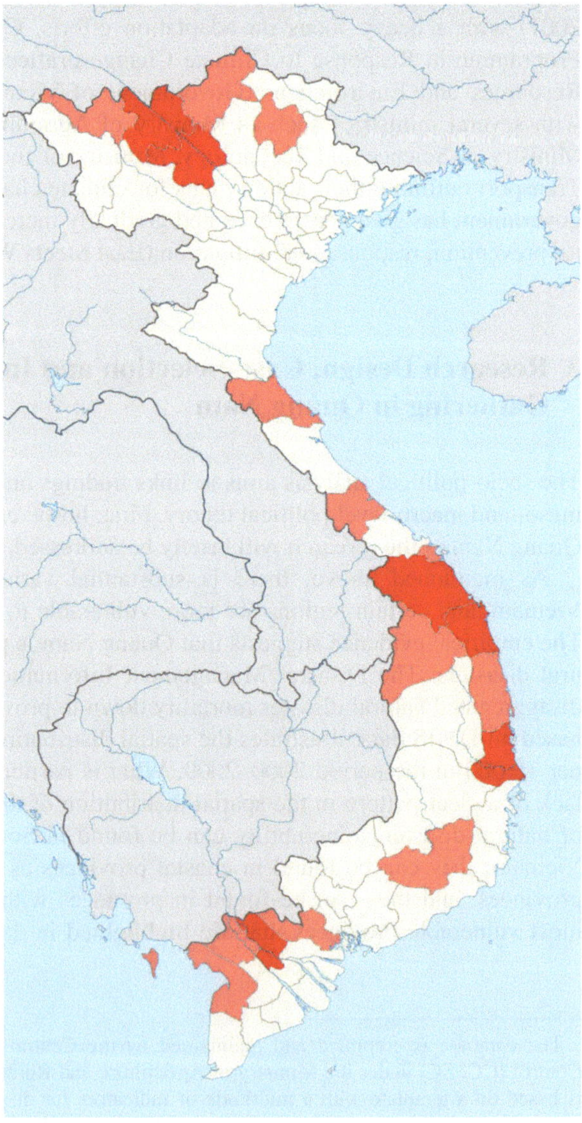

Fig. 2 Provincial map of disaster fatality intensity, 2000–2009. *Source* DIMS 2011, *dark red*: more than 2 deaths per 10,000 in the period 2000–2009, *light red*: more than 1 death per 10,000 in the period 2000–2009, *white*: less than 1 death per 10,000 in the period 2000–2009

respectively). Still, the six provinces carry losses that are three times as high compared to the rest of Vietnam. Of the six vulnerable provinces Quang Nam stands out as suffering from both a substantial natural disaster fatality rate (3 deaths per 10,000) and a high level of economic losses (5 million dong per capita). Figure 3 plots in the provinces according to their disaster fatality rates (x-axis) and disaster damages in million dong per capita (y-axis). The urban Da Nang municipality suffers most severely from disaster damages while the largely rural and agricultural dependent province of Quang Ngai is marred by the highest disaster mortality rate of more than 4/10,000. Still, Quang Nam scores high on *both* dimensions making it the most vulnerable province.

There is thus substantial variation between provinces with Quang Nam as one of the most vulnerable provinces in terms of fatalities and economic damages. This high level of vulnerability makes an analysis of Quang Nam particularly interesting as an extreme case. As such it not only constitutes an example of what might be in store for the whole of Vietnam with increasing climate variability in the decades to come, but it also provides a rich empirical case in which to analyze the socio-political impediments to effective disaster protection.

The analysis is based both on my own fieldwork conducted in Quang Nam in March 2011 as well as the work carried out by some of the other authors in this book. Most notably, the analysis will draw on the project's 2011 quantitative survey based on 166 households covering two coastal district (Hoi An and Nui Thanh), two intermediate level districts (Que Son and Dai Loc) and a mountainous district (Bac Tra My) (see Chapter 1). I conducted a total of 18 in-depth interviews across four districts within the Quang Nam province: Dien Ban, Hoi An, Duy Xuyen and Dai Loc. The majority of the interviews were with vulnerable

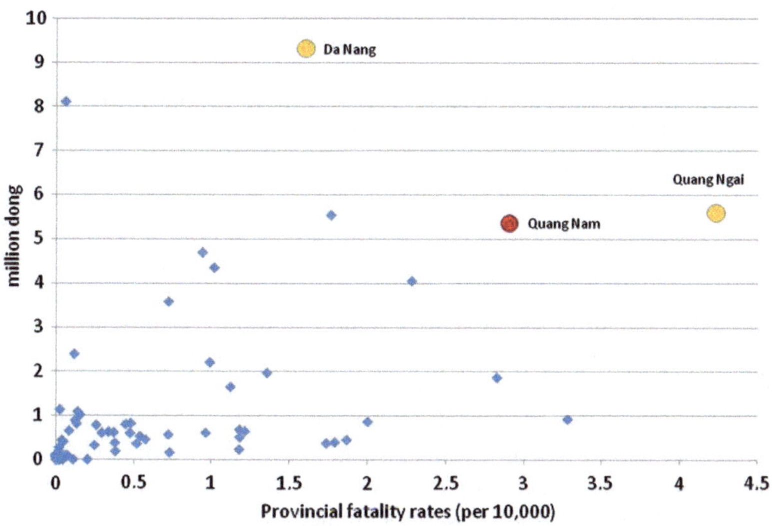

Fig. 3 Disaster damages and fatalities across provinces, 2000–2009. *Source* DMIS 2011

people that were regularly affected by flooding. The interviews were unstructured, open-ended (following Chambers 1997, p. 117), and often retrospective querying about the responses at the latest flooding.

Four public officials were also interviewed based on a core set of standardized questions. The purpose was to triangulate the quantitative information drawn from the project's survey with own observations and information gathered from interviews. The findings that emerged were subsequently related to meso- and macro-level political theory.

4 Socio-Political Impediments to Effective Disaster Protection

The following section will analyze some of Vietnam's socio-political impediments to effective disaster protection—not by assessing specific disaster policy designs and implementations but by integrating findings from Quang Nam with institutional and political theories at the meso- and macro-level.

It is the argument of this chapter that several of the adverse dynamics that have been identified is likely to be rooted in Vietnam's specific socio-political institutions. As indicated above, five deeper rooted political impediments to effective disaster management have been identified: (1) the lack of a social contract between the Vietnamese citizens and the government in disaster protection; (2) the lack of electoral competition both nationally but also locally; (3) the lack of devolution of power in local disaster management; (4) the technocratic bias in disaster management and climate change research; and (5) the weak social asset-base at the local level. There could very well be additional impediments to effective disaster protections; an obvious institutional barrier to effective public policy in Vietnam is corruption. The adverse effects of corruption on public administration and policy are well-documented and widely acknowledged by the major international organizations (UNDP 2008; Kaufmann and Kraay 2008; OECD 2010). A recent report published by the UNDP (2011) specifically addressed corruption in climate change adaptation, and concluded that corruption has the potential to seriously undermine adaptation efforts by diverting limited resources away from intended beneficiaries, inciting maladaptive activities and reducing the speed, scale and effectiveness of adaptation (UNDP 2011, p. 12). According to Transparency International, Vietnam ranks amongst the more corrupt countries with a score of 2.9 on a scale from 0 (highly corrupt) to 10 (very clean) in the perception based corruption index (Transparency International 2011). Several country-specific studies also point to the adverse effects of corruption on Vietnam's public policies (Danish Embassy, Swedish Embassy and World Bank 2011; Gillespie 2002; McCarty 2001). However, the topic of corruption did not emerge from the conducted fieldwork, and our survey was not attuned to cover corruption issues in any great depth. Hence, the chapter focuses on the five impediments above that can be directly related to fieldwork findings.

4.1 The Lack of a Social Contract in Disaster Protection

One of the main qualitative findings was that vulnerable households did not perceive disaster protection from the state as a right. In interviews, the respondents expressed hope that they would receive immediate relief (often in terms of rice and instant noodles) in the event of a disaster; they further hoped to receive some monetary compensation for the destruction caused by floods or storms; and they hoped soon to be eligible for new and more disaster-resilient housing. However, the immediate help during disasters (evacuation, food, shelter and water) as well as the rebuilding efforts (compensation and rebuilding of physical infrastructure) were clearly not based on ideas of state obligation and citizen's rights. While those who did not receive disaster assistance from the state certainly complained to us about what they considered to be arbitrary and unjustly distributed disaster assistance, they had not pushed their complaints through official channels (See also "Climate Change, Adaptation and the Environment in Central Vietnam").

Many vulnerable households did not appear to pay much attention to the source of the disaster relief. When respondents were asked about the disaster relief they had received, many did not qualitatively distinguish between relief aid from NGOs and aid from the state. A substantial number of vulnerable households were oblivious to whether the disaster relief they had received came from the state on one side or from domestic NGOs (for instance the Buddhist Associations) or international NGOs (for instance Oxfam and Red Cross) on the other. The fact that assistance from the state was not perceived to be qualitatively different from that of other sources is a direct consequence of most NGOs having to operate through state institutions.

Drawing on the work of Sen (1999, 2005, 2009), De Waal (1997, 2000), this absence of a right-based fundament to disaster management could be a consequence of a weak social contract between the Vietnamese citizens and the government. De Waal (1997, 2000) defines a social contract as 'the result of a popular movement successfully articulating a new right, and forcing a reluctant government to comply with its claims' (De Waal 1997, p. 11). The extent to which an issue is politicized depends both on the historical context as well as the dominant political institutions. According to De Waal, social contracts thrive best in political systems allowing for some kind of representation. This is because a contract requires two parties (in this case the state and the citizens) as well as the means to hold the parties accountable for the contractual obligations that have been laid out. Many democratic institutions (most notably competitive elections, the free press, and an independent judiciary) facilitate such accountability. As De Waal argues with respect to famine contracts: 'In conclusion, there can be anti-famine commitments and anti-famine programmes in the absence of democratic accountability, but an anti-famine contract requires the interested party—the people—to have some capacity to enforce the bargain' (De Waal 2000, p. 8).

Natural disasters have not, historically, been a salient issue in Vietnam. Even the largest recorded natural disaster, the 1971 flooding of the Red River causing fatalities in the vicinity of 100,000 (NOAA 2008) is limited when compared to the country's war fatalities numbering more than half a million in the Indochina War and more than a million in the Vietnam War. The fight for independence and the various anti-colonial movements were not related to natural disasters but had other ideological and military origins. Contrast this to India, where a widespread drought in 1943, and the failure of the British government to muster sufficient disaster protection, strengthened the nationalist movement that turned the ensuing famine into a basis for mass mobilization. The momentum for the nationalist movement led to independence already in 1947. The consequences of the British failure in 1943 together with the rhetoric of the nationalist movement created a strong incentive for the successive governments to avoid similar situations. A social contract of famine protection in drought situations emerged based on broad-based recognition by all major actors: the government, bureaucracy, civil leaders and the vulnerable segment of the population themselves.

Today, the right to food in India is enshrined in national law, and several national food security programs with universal suffrage are in place (the National Rural Employment Guarantee Act, the Integrated Child Development Services, Mid-Day Meals Scheme, and the Public Distribution System). Compare this to Vietnam, where—despite having ratified several international treaties concerning food safety in times of distress—Vietnam's legislative framework does not include a national law on the right to food, on food security or on nutrition (Xuan et al. 2009). Vietnam does not have specific judicial procedures to deal with the violations of rights to food, and it does not have a separate, independent institution to monitor the realization of the right to food in the country. In addition, Vietnam does not have a food-based national program to regularly provide free or subsidized food for the food-insecure people, instead the government supplies immediate food in emergency cases (Xuan et al. 2009). In fact, government policies on natural disaster preparedness and mitigation of natural disasters are mostly in the form of decisions or decrees and so forth. The value of these decrees is further undermined by a low level of community awareness. A report by East Meets West (2009) examined community awareness and concluded that citizens (and even government personnel) in Quang Nam's rural communities displayed limited knowledge of the basic government disaster policies.

4.2 The Lack of Electoral Competition

Even in the absence of a social contract, a pluralistic political system characterized by a competition for votes should facilitate more effective disaster relief. Political theory suggests that government responsiveness in times of distress is stronger (ceteris paribus) in more pluralistic political systems (Gasper and Reeves 2011; Morrow et. al. 2008; Boin et. al. 2005, 2008 ; Diamond 2008; Mesquita et. al. 2003; Sen 1999, 2005, 2009). Nobel Laurate Sen (1999, 2005,

2009) has famously asserted that democracy is the most effective remedy to famine. The argument is that in a democracy the government is faced with opposition parties, a free press and regular elections. This implies, according to Sen, that an elected government will be more responsive to popular demand, because voters have the power to replace it after the next election. This understanding of political dynamics is in line with Bueno de Mesquita et. al.'s influential monograph on political survival (2003). The authors argue that the politics of political survival is the essence of politics, and political leaders are assumed to be primarily interested in maximizing their survival time in office, and secondarily in maximizing their "income" from office. States are thus governed by a tenure-maximizing leadership that relies on a coalition of supporters to remain in power. Patron-client politics is an inherent part of the logic of political survival, as political loyalty is given in return for preferential access to public resources or in return for the distribution of public goods. In many settings, political parties as well as ethnic and/or regional groups form the organisational bases for patronage networks, since members of these groups will often already share loyalty ties with the political leaders (World Bank 2009). Patronage networks are particularly important in situations of crisis where time is of the essence and fiscal resources are limited.

Hence, government responsiveness is not linked to the benevolence of a democratic regime vis à vis an authoritarian regime, but to the fact that governments—fighting for survival (re-election)—will go to great lengths to avoid a humanitarian catastrophe. It is a political responsiveness mechanism whereby the government responds to pressure exerted by the opposition, the media and the general public by undertaking disaster protection in times of distress. The media carries two important functions: one is as a mediator of information in times of crisis and one as a watchdog making the government accountable for the disaster management.

Both Amartya Sen and Alex De Waal argue that although socialist regimes might be very good at generating huge improvements in people's livelihoods, the protection of livelihoods and disaster relief is perceived to be a state-endowment rather than a citizen right. Nowhere was this clearer than during the largest famine in history (with 30 million fatalities), namely the Chinese famine during 1959–1961. Prior to the famine, impressive strides had been made with regards to malnutrition and infant mortality, but the commitment to help the vulnerable was based on a top-down initiated ideology—not a social contract between the government and its citizens. In normal years this mattered little; in times of distress, however, the government was able to ignore the suffering of millions (for the greater good, of course) with devastating consequences.

Like China, Vietnam has succeeded in improving the quality of life for many of its citizens, and thereby effectively lowered their vulnerability to disasters. However, the political theories of both democratic response and social contract draw attention to the fragility of the relationship between the government of Vietnam and its citizens; the relationship cannot be characterized as symbiotic fortified by social contracts and a pluralistic system of accountability.

4.3 Decentralization Without Devolution

One of the likely implications of a highly centralized, authoritarian system is that provinces and communes will have limited autonomy in disaster management. While decentralization processes has accelerated in Vietnam during the last decades, the local political entities remain instruments for policy implementations (Mattner 2004; Fforde 2003). The decentralization processes were not followed by true devolution of power neither politically nor fiscally, and many scholars point to the lack of substantial devolution of power in the wake of Vietnam's decentralization. In his governance analysis of Vietnam, Vo (2009) concludes that Vietnam's subnational governments remain very weak. He describes how the national government continues to set tax bases and tax rates; how the tax administration is nationally centralized; and how there is a heavy interference on sub-national spending programs from the national government (Vo 2009, p. 417). This is supported by Nguyen and Anwar's (2011, p. 6) recent analysis of decentralization in Vietnam highlighting that 'the government has a big say in spending decisions taken by each of the sub-national governments, and hence the level of autonomy is severely restricted'. The Vietnamese government strives to take the governmental agencies as close as possible to the people, to the localities, to the villages, in order to consolidate their organizational base in the rural areas (especially in the South). Decentralization is less about *devolution* of power as it is about the *exercise* of power. This drives some of the impediments to effective disaster protection that has been observed in Quang Nam.

One of the few consistent findings in the disaster literature is the fact that the link between the community and local governance is vital in a disaster situation. In situations where all community members face similar kinds of covariate risks, the local government institutions can play an important role as a protector of the last resort. The 2010 World Development Report explicitly acknowledge that local communities typically precede national government in climate change adaptation (World Bank 2009, p. 20), Agrawal (2008) has documented the importance of local institutions in livelihood adaptation based on 118 cases of good-practice adaptation from around the globe. In 77 of the cases, the primary structuring influence for adaptation flows from local institutions, and in all cases of external support, the support is channeled through local formal and informal institutions. Adger also emphasizes the importance of generating synergy between the community and the local state institutions. He notes that state-society linkages are important both for wider sustainable development and for the co-management of resources. 'States can facilitate sustainable and resilient resource management and enhance adaptive capacity' (Adger 2003, p. 395).

More concretely with respect to Vietnam, Shaw (2006) recommends stronger integration between communities and government in Vietnam's climate change adaptation efforts, and concludes (Shaw 2006, p. 17) that 'self-governance and local governance are the key factors' in ensuring such policy integration. One of Adger's main conclusions from his many analyses of environmental change and social resilience, in particular his analysis of communities in coastal Vietnam, relates precisely to the importance of generating synergy between community knowledge and characteristics on one side and local state institutions on the other (Adger 1999).

The implications of this lack of devolution to the local institutions clearly emerged from the survey. Despite the somewhat loaded question of whether climate and weather issues had ever been discussed in local meetings, less than half of the respondent felt in a position to answer the question. Among the respondent who answered, the overwhelming majority (more than 90 %) described meetings that could best be characterized as fora for state announcements and information dissemination. People were in their own words 'reminded', 'informed', 'encouraged' and physically 'moved.' Only 6 % of the responded actually included the same verb as the question itself, namely that meeting included 'discussions.' Thus, while there are many voluntary organizations at the community level in Quang Nam (as well as in Vietnam in general) that play a key role in disaster management, they are top-led organizations that function as an extension of the Communist party. Bruun reaches the same conclusion in "Rural Households: Socio-Economic Characteristics, Community Organizing and Adaptation Abilities" where he cautions against understanding 'communities' as something distinct from the government. In the common perception, communities are understood in the context of state institutions rather than civil societies. Hence, the Communist party organization stretches downwards through the government tiers, and is still firmly in place and dominating all aspects of rural organization. The recent one trillion VND project on raising community awareness and disaster risk management, for instance, builds on clear channels of authority and expertise extending from the central level down to the provinces (Government of Vietnam 2009). This has adverse implications for the development of a community social asset-base; social assets cannot flourish in the shadows of a dominant and all-encompassing state.

4.4 Top-Down and Technocratic Disaster Protection

Vietnam's approach to natural disaster management is not only centre-driven but it also has a clear focus on gauging the biophysical impacts of climate change (UN 2009). Vietnam's disaster protection tends to concentrate on reducing physically determined vulnerability with limited attention to the socially determined vulnerability. Through a rigorous analysis of Vietnam's climate change strategy, Fortier (2010, pp. 38–240) concludes that climate change in Vietnam is primarily understood as a bio- and geophysical problem that can grasped by conducting increasingly complex scenario-analyses, and that can subsequently be addressed by implementing technical solutions. Climate change is viewed as an external (and unjustly imposed) threat that will result in some concrete natural hazards that can be forecasted and then dealt through dykes, higher bridges, water reservoirs, resilient infrastructure and so on—what is often referred to as 'climate proofing' (UN 2009). Bruun (2012) has also identified clear vertical structures and hierarchies in the Vietnamese climate change research and disaster adaptation policies that confine and standardize policies along the more technocratic dimensions of the main line ministries.

This technocratic understanding of natural disasters can not only be traced back to political institutions but also to the socialist ideology that is often based on technocratic approaches to societal problems. Natural disasters are by their very nature random and complex while the political dimension of a socialist ideology (if not the economic to the same extent any longer) builds on notions of order and natural laws. Attention to social assets in disaster management has therefore been scarce given the tendency to treat disaster management mostly as an engineering issue calling for technical solutions. Local meetings with communes, for instance, are often used by the local government to share weather forecasts, and while this information should surely an integral part of any disaster management strategy, it should be but one part of local government and community interaction.

4.5 Weak Social Assets at the Community and Hamlet Levels

There is little doubt that community social capital matters in local disaster management (Rubin and Rossing 2012; Pelling 2011; Koh and Cadigan 2008; Nakagawa and Shaw 2004). However, the Vietnamese government's centre-driven and technocratic approach to natural disaster management undermined the scope for community social assets and participation.

My interviews with vulnerable households did not only indicate a substantial reliance on the state for disaster protection, but they also revealed a weak reliance on local social assets in disaster protection. A few vulnerable households stated that village/neighbor-relations played a role in their disaster recovery (a family even told me that a wealthier family had paid for the reconstruction of their house) but the majority depended on the state and close family ties. To the extent that private donations did take place, they were often channeled through the local government. Even in the few cases where families did emphasize disaster protection based neighbor-relations, they appeared to refer to ad hoc donations rather than local level cooperation and reciprocity. These findings are confirmed by the survey. When respondents were asked to identify the most important sources of economic assistance when coping with natural disasters or other kinds of distress, 73 % of the respondents highlighted relief from the state. This is indicative of a strong reliance on the state (14 % answered that they had received relief from NGOs; a single household had only received relief from NGOs). Equally noteworthy was the fact that only 4 % of the respondent identified neighbors as an important (not even the most important) source for disaster protection (23 % of the respondents had either not received disaster relief or chose not to answer the question). This hints at an underdeveloped social asset-base at the commune level. Contrast this, for instance, to the vulnerable communities in Mexico's southern state of Chiapas, where a study asked respondents to rate different possible sources of financing for disaster relief by scoring them between one (irrelevant) and five (very important) (Saldaña 2006). The study found that in times of stress community members relied heavily on social assets: relatives living within the community were scored to be the most important source of disaster finance with neighbor solidarity coming in third.

The Quang Nam respondents were queried in the survey whether their community was actively preparing for natural disasters. The rationale behind asking this question, other than receiving information on natural disaster management, was how the concept of 'the community' was interpreted among the respondents: did they associate it with civil society or the state? Supposedly, Vietnamese usually associate the term 'community' directly with state structures. This assumption was supported by the survey answers. A small minority (12 %) of the respondent emphasized disaster management that could be rooted in civil society such as coordinating local meetings, streamlining village organization, improving early information systems and rescue plans, and a reliance on family networks. The vast majority of the respondents (62 %), however, automatically related the question to state-led disaster policies such as the provision of better roads, drainage, dikes/dams and housing (26 % either abstained from answering or answered that the community was not preparing).

Another recent survey from Quang Nam conducted by East Meets West Foundation (2009) reinforced many of the same findings that have already been presented in this chapter. East Meets West selected 25 local communities to represent the most vulnerable regions that are heavily and regularly affected by severe weather and natural disasters. A total of 125 households were surveyed on multiple dimensions of climate change adaptation where we shall focus on the households' relationship to the government. According to the survey, communities were often unaware of national policies and procedures related to disaster preparedness and planning. Community interviews showed that about 49 % of total interviewees were unaware of national policies and procedures, and 72 % of interviewees did not know about disaster plans that were made at the community or commune level (East Meets West 2009, p. 24). In the 25 communities surveyed, there had been no field exercises dealing with natural disaster preparedness. The information collected in the interviews revealed that only 37 % of the interviewees were aware of training sessions that had recently been held by the local authorities on disaster preparedness at the commune or community level.

5 Conclusion

This chapter assessed Vietnam's vulnerability to natural disasters, and presented evidence from the Quang Nam province of adverse socio-political dynamics and structures that might exacerbate these vulnerabilities. Vietnam suffers from relatively high incidents of natural disasters due to the country's high susceptibility to climate-induced natural hazards. Quang Nam was identified as being particularly vulnerable to natural disasters along the dimensions of economic damages and fatalities. The province therefore constituted a purposely selected case to conduct the socio-political analysis. The fieldwork from Quang Nam indicated that vulnerable households did not perceive of disaster protection as a right, and that they did not consider disaster protection as a legitimate demand on the state.

According to the political theories of both Sen (1999), De Waal (2000), the lack of pluralistic political institutions as well as the absence of a social contract of disaster protection makes it difficult to hold the government accountable for belated or inadequate disaster protection. The state could still implement numerous policies that increase the welfare of its citizens; however, there is no *obligation* on the state to provide protection during times of distress. The institutional consequence is that many rights (most notably the right to food and water) are not formalized in national law; instead emergency relief is handed out ad hoc during natural disasters.

The empirical evidence also suggested that although the vulnerable respondents did not perceive disasters protection as a state *obligation,* they still repeatedly emphasized their dependence on the state for protection in times of crisis. The fact that many vulnerable households relied mostly on state support during and in the immediate aftermath of a disaster appears to be indicative of (and might further erode) an underdeveloped social asset-base at the community level. The several voluntary organizations that exist at the community level (some of which play a vital role in disaster management) are top-led organizations deeply integrated with the Communist Party of Vietnam. Most NGOs (with the Buddhist Associations as a noticeable exception) are also obligated to work in close liaison with state institutions.

While decentralization processes has accelerated in Vietnam during the last decades, the local political entities remain instruments for policy implementation. Thus, the Vietnamese decentralization process has not been followed by a true devolution of fiscal and political power. The relationship between communes and local government institutions could not, therefore, be described as symbiotic but instead established and controlled by the central government. Furthermore, disaster research and management in Vietnam appeared to be biased towards natural sciences and preoccupied with technical solutions to climate change variability. This leaves very little scope for community participation and local knowledge sharing—a vital ingredient, research has shown, in bolstering disaster resilience.

Disaster management meetings were used to disseminate instructions and information from above; it was less a participatory process where community assets could be strengthened in order to mitigate the impacts of natural disasters. Designing and implementing the right policies for disaster management is an important dimension of disaster resilience. However, a more fundamental shift in the disaster management paradigm in the light of the future implications of climate change would demand a more fundamental shift in the political structures and institutions.

References

Adger WN (1999) Social vulnerability to climate change and extremes in coastal vietnam. World Dev 27(2):249–269
Adger WN (2003) Social capital, collective action and adaptation to climate change. Econ Geogr 79(4):387–404
Adger WN (2006) Vulnerability. Glob Environ Change 16(3):268–281
Agrawal A (2008) The role of local institutions in adaptation to climate change. Paper presented at workshop on social dimensions of climate change. World Bank, Washington DC
Asian Development Bank (2009) The economics of climate change in Southeast Asia: a regional review. Jakarta. http://www.adb.org/Documents/Books/Economics-Climate-Change-SEA/PDF/Economics-Climate-Change.pdf
Boin A, Mcconnell A, Hart P (2008) Governing after crisis—the politics of investigation, accountability and learning. Cambridge University Press, Cambridge
Boin A, Hart P, Stern E, Sundelius B (2005) The politics of crisis management: public leadership under pressure. Cambridge University Press, Cambridge
Bruun O (2012) Sending the right bill to the right people: climate change, environmental degradation and social vulnerabilities in Central Vietnam. Weather Clim Soc 4(4):250–262
Chambers R (1997) Whose reality counts? putting the first last. ITDG, London
Cuong H, Yen P, Nga N, Thuy H, Ha H (2009) Overview of climate change scenarios in the World, in Vietnam and mid-central Vietnam. Sub-project 2: assessing effects of and responses to climate change on environment and socio-economic development in mid-Central Vietnam. Vietnam Academy of Science and Technology, Institute of Geography (www.ecoenvi.org)
Danish Embassy, Swedish Embassy and the World Bank (2011) Recognizing and reducing corruption risks in land management in Vietnam. National Political Publishing House, Hanoi
Dasgupta S, Laplante B, Meisner C, Wheeler D, Yan J (2007) The impact of sea level rise on developing countries: a comparative analysis. World Bank Policy Research Working Paper 4136. http://go.worldbank.org/XU9B5UFR30
De Waal A (1997). Famine crimes, Villiers Publication
De Waal A (2000). Democratic political process and the fight against famine. IDS Working Paper 107
Diamond L (2008) The spirit of democracy: the struggle to build free societies throughout the World. Times Books, New York
DMIS (2011) Disaster management information system, Central Committee for Flood and Storm Control. http://www.ccfsc.gov.vn/dmis/Reports.aspx?cat=0
East Meets West (2009) Climate change adaption survey http://downloads.eastmeetswest.org/pubs/emw_climatechangeadaptsurvey.pdf
EM-DAT (2012) Emergency events database. http://www.emdat.be/
Fforde A (2003) Decentralization in Vietnam—working effectively at provincial and local government level—a comparative analysis of long an and Quang Ngai provinces. Australian Agency of International Development
Fortier F (2010) Taking a climate chance: a procedural critique of Vietnam's climate change strategy. Asia Pacific Viewpoint 51(3):229–247
Gasper J, Reeves A (2011) Make it rain: retrospection and the attentive electorate in the context of natural disasters. Am J Polit Sci 55(2):340–355
General Statistics Office of Vietnam (2009) Statistical yearbook 2009. http://www.gso.gov.vn/default_en.aspx?tabid=515&idmid=5&ItemID=10439
Gillespie J (2002) Chapter 10—the political-legal culture of anti-corruption reforms in Vietnam. In: Lindsey T, Dick HW (eds) Corruption in Asia: rethinking the governance paradigm, pp 167–200
Government of Vietnam (2007) National strategy for natural disaster prevention, response and mitigation to 2020. http://sheltercentre.org/sites/default/files/vietnam-national_strategy.pdf
Government of Vietnam (2008) National target program to respond to climate change. Unofficial translation. http://www.isgmard.org.vn/VHDocs/NationalPrograms/NTP%20RespondtoClimateChange.pdf

Government of Vietnam (2009) Decision to approve the project: community awareness raising and community-based disaster risk management (CBDRM). http://www.ccfsc.gov.vn/resources/users/6D696775656C/CBDRM%20Decision%201002.pdf
Hodell DA, Curtis J, Brenner M (1995) Possible role of climate in the collapse of the classic maya civilization. Nature 375:391–394
IPCC (2007) Climate change 2007: impacts, adaptation and vulnerability. Contribution of working group II to the fourth assessment report of the IPCC. IPCC, Geneva. http://www.ipcc.ch/publications_and_data/ar4/wg2/en/contents.html
Kahn M (2005) The death toll from natural disasters: the role of income, geography, and institutions. Rev Econ Stat 87(2):271–284
Kaufmann D, Kraay A (2008) Governance indicators: where are we, where should we be going? http://siteresources.worldbank.org/INTWBIGOVANTCOR/Resources/KKGovernanceIndicatorsSurveyWBROSpring2008.pdf
Koh H, Cadigan R (2008) Chapter 13—disaster preparedness and social capital. In: Kawachi I, Subramanian SV, Daniel K (eds) Social capital and health. Springer, New York
Mattner M (2004) Power to the people? local governance and politics in Vietnam. Environ Urbanization 16:121–128
Mesquita B, Smith A, Siverson R, Morrow J (2003) The logic of political survival. Cambridge, Mass, MIT
Nakagawa Y, Shaw R (2004) Social Capital: A Missing Link to Disaster Recovery. International Journal of Mass Emergencies and Disasters 22(1):5–34
National Oceanic and Atmospheric Administration (2008) NOAA's top global weather, water and climate events of the 20th century. http://www.noaanews.noaa.gov/stories/images/global.pdf
Nguyen L, Anwar S (2011) Fiscal decentralisation and economic growth in Vietnam. J Asia Pacific Econ 16(1):3–14
McCarty A (2001) Governance institutions and incentive structures in Vietnam. In Building Institutional Capacity Building in Asia Project commissioned by the Ministry of Finance, Japan, pp 26–62
Morrow J, Mesquita B, Siverson R, Smith A (2008) Retesting selectorate theory: separating the effects of W from other elements of democracy. Am Polit Sci Rev 102(3):393–400
OECD (2010) Development co-operation report 2010 http://www.cociscampagnapace.it/materiali/OECD2010.pdf
Pelling M (2011) Adaptation to climate change—from resilience to transformation. Routledge, Abington
Quy N (2011) Assessing impacts of climate change on disaster related stream flow (flood and drought) in Quangnam, Vietnam. Sub-project 5: Assessing effects of and responses to climate change on environment and socio-economic development in mid-central Vietnam. Vietnam Academy of Science and Technology, Institute of Geography (www.ecoenvi.org)
Roberts J, Parks B (2007) A Climate of Injustice – Global Inequality, North South politics, and Climate Policy. Cambridge, Mass., MIT Press
Rossing T, Rubin O, Brisson I (2010) Building short-term coping capacity and longer-term resilience through asset-based adaptation. In: Verner D (ed) Reducing poverty, protecting livelihoods, and building assets in a changing climate. World Bank Publication, Washington, pp 267–303
Rubin O (2012) Socio-economic determinants of vulnerability to climate-induced natural disasters—cross-provincial evidence from Vietnam
Rubin O, Rossing T (2012) National and local vulnerability to climate-related disasters in Latin America: the role of social asset-based adaptation. Bull Latin Am Res 31(1):19–35
Saldaña-Zorrilla S (2006) Stakeholders' views in reducing rural vulnerability to natural disasters in southern Mexico: hazard exposure, coping and adaptive capacity. http://www.start.org/Program/advanced_institute3_web/Final%20Papers/SALDANA%20FINAL%20PAPER%20AIVGEC.pdf
Satterthwaite D, Huq S, Reid H, Pelling M, Lankao P (2007) Adapting to climate change in urban areas: the possibilities and constraints in low and middle income nations. London, International Institute for Environment and Development (IIED) http://pubs.iied.org/pdfs/10549IIED.pdf

Sen A (1999) Development as freedom. Knopf, New York

Sen A (2005) The argumentative Indian: writings on Indian history, culture and identity. Allen Lane, London

Sen A (2009) The idea of justice. Allen Lane, London

Shaw R (2006) Community-based climate change adaptation in Vietnam: interlinkages of environment, disaster, and human security. In Sonak S (ed) Multiple dimension of global environmental changes. New Delhi, TERI Publication http://unfccc.int/files/adaptation/application/pdf/nwpexpert_shaw_2006.pdf

Simms A, Reid H (2006) Up in smoke? Latin America and the caribbean: the threat from climate change to the environment and human development: the third report from the working group on climate change and development. http://pubs.iied.org/10017IIED.html

Transparency International (2011) Corruption perception index http://cpi.transparency.org/cpi2011/results/

UN (2009) Viet Nam and climate change: a discussion paper on policies for sustainable human development. http://un.org.vn/en/publications/un-wide-publications/doc_details/108-viet-nam-and-climate-change-a-discussion-paper-on-policies-for-sustainable-human-development.html

UNDP (2004) Reducing disaster risk. A Challenge for development—a global report. http://www.undp.org/cpr/whats_new/rdr_english.pdf

UNDP (2008) Corruption and development: a primer http://www.undp.org/content/dam/aplaws/publication/en/publications/democratic-governance/dg-publications-for-website/corruption/Corruption_and_Development_Primer_2008.pdf

UNDP (2011) Staying on track: tackling corruption risks in climate change http://www.undp.org/content/dam/aplaws/publication/en/publications/democratic-governance/dg-publications-for-website/staying-on-track–tackling-corruption-risks-in-climate-change/Staying_on_Track_corruption_risk_in_CC.pdf

Viner D, Bouwer L (2006) Linking climate change adaptation and disaster risk management for sustainable poverty reduction—Vietnam country study. European Commission, Vulnerability and Adaptation Resource Group http://www.gcca.eu/usr/documents/EC-VARG-VietnamStudy_20101071657 47.pdf

Vo D (2009) Fiscal decentralisation in Vietnam: lessons from selected Asian nations. J Asia Pacific Econ 14(4):399–419

WDI (2012) World development indicators. World Bank. http://data.worldbank.org/

Wisner B, Blaikie P, Cannon T, Davis I (2004) At Risk. Routledge, Abington

World Bank (2009) World development report 2010: development and climate change. http://go.worldbank.org/ZXULQ9SCC0

Xuan T, Thanh H, Phuong D, Thuan H, Son P, Giang D (2009) Viet Nam: a right to food assessment. Hanoi, ActionAid http://www.actionaid.org.vn/Modules/CMS/Upload/Download/Vietnam-RtF-Assessment.pdf

Livelihood Strategies Under the Constraints of Climate Change Vulnerability in Quang Nam

Thorkil Casse

Abstract This chapter examines how vulnerability can be measured in quantitative terms. By using the Ketsana typhoon's devastating effects on livelihoods as an example, the chapter shows that households whose livelihoods are based on economic activities like acacia production and shrimp farming suffered the most. A survey of 166 households showed that 88 % had experienced substantial losses. In this case study, the degree of severity of the losses is not an useful indicator of vulnerability in Quang Nam. The people who were most aware about climate changes were also the ones most satisfied with their district's disaster plans, and these were better covered by disaster relief schemes. At the other end of the scale, we found a district where only half of the households were satisfied with their district's disaster plans, and half of the total claimed they did not know anything and had not received any disaster relief. In our study area, knowledge of climate change, satisfaction with local disaster planning and receipt of disaster relief seem to go hand in hand.

Keywords Vulnerability • Livelihood strategies • Costs of natural disasters • Off-farm activities • Disaster compensation

1 Introduction

According to the IPCC (2007), Vietnam will experience more, and increasingly severe, natural disasters as a result of climate change. Specific studies undertaken in Quang Nam have shown that temperatures have increased by a factor of 0.01 in the period 1971–2000—a difference of 0.5 °C (Van Tan 2010, p. 45). Figures on changes in precipitation show an equally low degree of variation from 1981–1990 to 2001–2008. From the first period to the second, only one weather station out of eight in Quang Nam showed an increase in average yearly heavy rain days

T. Casse (✉)
Department of Environmental, Social and Spatial Change, Roskilde University, Roskilde, Denmark
e-mail: casse@ruc.dk

O. Bruun and T. Casse (eds.), *On the Frontiers of Climate and Environmental Change*, Environmental Science and Engineering, DOI: 10.1007/978-3-642-35804-3_7,

exceeding 2 days recorded (Hoang et al. 2010). Nevertheless, we have seen an increase in tropical typhoons which are building up in strength. Typhoons producing a wind level of 12 or more have increased from two in each of the decades (1980–1990 and 1991–2000) to five in the period of 2000–2008 (Van Tan 2010, p. 35). The major rainy season in Central Vietnam can be attributed to cyclone activity which is responsible for heavy rainfall and consequent flood events (Chen et al. 2011). The population in Quang Nam will face the full force of the natural disasters, even if we cannot establish the relationship between the change in temperature and the number of natural disasters with any certainty.

This study aims at fulfilling three objectives:

1. To describe the concept of climate change vulnerability in economic, social and institutional terms
2. To discuss the livelihood strategies of the income groups
3. To discuss the local people's assessment of local authorities' disaster management plans.

The observations presented in the chapter draw on findings from a household survey conducted in five districts during 2009–2010. The chapter shows a significant difference between the various income groups in how disasters, using the example of typhoon Ketsana, have affected their livelihoods. Houses of poor families are of inferior quality, and are heavily damaged by typhoons. Rich and middle income families see the economic activities being curbed by the destruction of typhoons. A difference between the local people's perception of district disasters plans is notable, despite the official policy of offering all districts offering similar disaster reduction schemes. Finally, 20 % of all households did not receive any disaster relief from the government or NGOs. Once again, the official view holds that all households suffering from the damaging effects of typhoons are eligible for aid relief from the government (see "Impediments to Climate-Induced Disaster Management: Evidence from Quang Nam Province, Central Vietnam"). The disaster relief did not amount to more than 1 million Dong, so major investments in plantations or shrimp farming were all lost.

2 Impact of Disaster Losses on Different Income Groups

2.1 Concepts of Vulnerability

Vulnerability and poverty measures are closely related, though they are often given different meanings by different authors. Moser (1998, p. 3) observers that 'although poor people are usually among the most vulnerable, not all vulnerable people are poor'. Chaudhuri et al. (2002), argue that vulnerability can be seen as the risk of a certain household becoming poor, whereas poverty is the ex-post measurement of the income level (or broader well-being) of the same household.

The Intergovernmental Panel on Climate Change's (IPCC) Third Assessment Report (TAR) defined climate change vulnerability to be a function of exposure, sensitivity and adaptability. Kelly and Adger (2000) argue that the poor suffer not only from the event of the disaster itself but to an equal extent from the social conditions under which they live. They define vulnerability 'in terms of individuals, and social groupings to respond to, in the sense of cope with, recover from or adapt to, any external stress placed on the livelihoods and well-being' (ibid. 328). They highlight the presence of inequality which can augment the collective vulnerability of a community.

In addition to poverty and inequality, institutional adaptation ought to be included in a vulnerability analysis. Building on the views of Kelly and Adger, O'Brien et al. (2008) distinguish between end-point and starting point vulnerability. In their view, end-point vulnerability is outcome vulnerability in linear terms, linking climate change projections to an impact on an exposure unit (for example a household). In other words, end-point vulnerability claims to be able to measure, more or less precisely, the impact on incomes and even buildings of any major change in temperatures, increase in rainfall or increase in the frequencies of natural disasters (depending on the climate change criteria). O'Brien et al. (2008) maintain that up until now, the end-point or outcome vulnerability has dominated the debate, and adaptation policies are defined quite narrowly. Starting point vulnerability is contextually dependent.

The ability to cope with the negative impacts of natural disasters is a function of various socio-economic factors. In a study of New Orleans, Masozera et al. (2007) list a number of factors, such as age, gender, land property, education and social dependency on financial transfers from outside as being the main factors which explain vulnerability in the aftermath of hurricane Katrina, which occurred in 2005. Dasgupta and Baschieri (2010) propose a logistic model to explain the risk of going from a reasonable functioning household to a more vulnerable income group, when rainfall decreases between the regions in Ghana.

A five component vulnerability definition is suggested by Cannon (2008), which differs from the Kelly and Adger (2000) by talking about two types of protection: self-protection (income and resources) and social protection (substitutes for self-protection). We found households in the study area needing to turn to the community to request for assistance, and this could be seen as an indication of social protection. Vulnerable households did not perceive disaster protection from the state as a right; an issue which is discussed further in "Impediments to Climate-Induced Disaster Management: Evidence from Quang Nam Province, Central Vietnam" in this book. Füssel and Klein (2006) distinguish between first and second generation vulnerability assessments; the latter include non-climatic drivers such as economic diversification, education and the strength of social network. Education is an additional indicator of possible presence of vulnerability we will examine in the following.

In this chapter, we will attempt to quantify the concept of vulnerability, based on the three tiers (poverty, inequality and institutional adaptation), as suggested by Kelly and Adger (2000). While discussing the institutional framework we will apply a similar approach to that of Dasgupta and Baschieri (2010). We work within the framework of end-point vulnerability, though we claim that only a part of vulnerability can be measured by this method.

2.2 Income Losses Between Different Income Groups

During the period November 2009 to February 2010, we collected information from 166 households in five districts (Hoi An, Dai Loc, Que Son, Nui Thanh, and Bac Tra My). The questions were related to general information about household size, economic activities, economic losses following the Ketsana typhoon and the household awareness of climate changes.

Different approaches exist for grouping households into income groups. One approach is the official Vietnamese threshold for nearly poor households of 400,000 Dong per head per month and the poor of 210,000 Dong per head per month (Republic of Vietnam 2011). On average, the number of persons in each household is 4.4 or a threshold of 17 million Dong per year per household. In the following we use 17 million Dong level as the threshold to classify a household as being poor, yielding the following household groupings: less than 17 million Dong (poor), between 17 and 40 million Dong (middle income), and more than 40 million Dong (rich). Many very poor households (closer to five million dong per household) are made up of retired persons, and significance between economic losses and type of economic activities is also very difficult to establish, since they do not engage in economic activities.

Table 1 shows the income groups by district. As can be deduced from the table, the distribution of income among the districts is fairly equal. Since our approach was to have all three income groups represented from the five districts in our survey, this table cannot be interpreted as an indication of the real income distribution by district. Table 2 indicates the percentage of disaster induced losses compared to total income by rich, middle income and poor households. As can be seen from the table, even in relative terms, the rich income households are those mostly affected by the losses incurred, following the devastating impacts of the typhoon Ketsana. We also

Table 1 Income groups by district

		District					Total
		Bac Tra My	Dai Loc	Hoi An	Nui Thanh	Que Son	
Income groups	Below 17 millions	19	10	13	21	17	80
	Between 17 and 40 millions	8	7	5	19	13	52
	Above 40 millions	5	7	7	8	7	34
Total		32	24	25	48	37	166

Source Household survey

Table 2 Income loss fractions by poor, middle income and rich households

		Income groups			Total
		Below 17 millions	Between 17 and 40 millions	Above 40 millions	
Loss fraction groups	0–10 %	9	5	0	14
	Between 10 and 50 %	15	2	0	17
	Above 50 %	56	45	34	135
Total		80	52	34	166

Source Household survey

Table 3 Correlation between log total losses and independent variables, poor households, less than 10 million dong[a]

	Unstandardized coefficients		Standardized coefficients		
Model	B	Std. error	Beta	t	Sig.
(Constant)	4.149	0.301		13.769	0.000
Small scale production	−0.001	0.000	−0.299	−2.291	0.026
House condition	−0.198	0.092	−0.279	−2.156	0.036
Land insufficiency	0.080	0.050	0.208	1.597	0.117

[a] Dependent variable: log total losses
[b] $R^2 = 0.15$
Source Household survey

note that ¾ of the sampled households were heavily affected by Ketsana (disaster losses amount to more than 50 % of total yearly income in 2008).

In the next section we look at regression equations, correlating the disaster losses to a variety of socio-economic factors for the three main income groups. Can we explain what factors might increase or decrease the vulnerability of the households in pure economic terms?

The following are the results for poor households ($n = 55$ or $n = 80$):

The poor households group is a very heterogeneous category. In the lower end ($n = 55$, less than 10 million dong in yearly income, Table 3), we could identify the type of house (concrete house or not, $p = 0.036$) and petty production ($p = 0.026$) as the two factors correlated with the disaster losses. In both cases, the coefficient is negative, suggesting that a concrete house and non-agricultural income lowers the economic vulnerability. At the higher end (a household income between 10 and 17 million dong), we find that agricultural income is close to significance ($p = 0.11$) and plantation area is significant ($p = 0.019$) with disaster losses. The characteristics of the high end of the poor household group resembles the pattern of the middle income group (see below). For both linear analyses, the strength of linear association is low (approximately 15 % of the correlation is explained), supporting the hypothesis of a fairly heterogeneous group.

The logarithmic value of the total disaster losses is correlated to small scale production, house condition (in this case the concrete houses of the poor households are very fragile to disaster effects) and land insufficiency (nearly statistically significant). Based on the household's reported income and losses figures, we do not find any correlation which is directly related to the economic activities of the very poor households. More off-farm activities, like small production, and better houses render the poor households less vulnerable in economic terms. The most important factor, not surprisingly, causing economic vulnerability of the poor households is lack of land.

The following are the results for middle income households (n = 77):

Compared to the poor households, the middle income households face increasing vulnerability due to their forest activities and income from farming. Attempting to include forest income yields no significant results. This is probably explained by the fact that earnings from forestry were low in 2009/2010, and the forest expansion (plantations) is to be seen as investments with no payoff as yet. Wage income is another off-farm economic activity reducing economic vulnerability (Table 4).

The results for richer households (n = 34) are:

For the rich households, the income from shrimp farming is significant at a 10 % rejection level. The option of having access to factory earnings decreases economic vulnerability. As in the case of middle income households, the forest area is significant or close to significance. No other factor proved significant for explaining the disaster losses of the rich households (Table 5).

The results of the regression tests indicate that the reported disaster losses are linked to the quality of houses or petty production for the lowest income households, or economic activities (agriculture and forestry) for the high end of the poor household group. Petty production tends to decrease the economic vulnerability of the poor households.

Table 4 Correlation between log total losses and independent variables, middle income households[a]

Model	Unstandardized coefficients		Standardized coefficients	t	Sig.
	B	Std. error	Beta		
(Constant)	2.360	0.808		2.920	0.005
Forest area	9.154E–6	0.000	0.225	1.900	0.063
Income from wage labour	−3.834E–5	0.000	−0.326	−2.739	0.008
Log agriculture income	0.378	0.213	0.212	1.777	0.081

[a] Dependent variable: log total losses
[b] $R^2 = 0.54$
Source Household survey

Table 5 Correlation between log total losses and independent variables, rich households[a]

Model		Unstandardized coefficients		Standardized coefficients	t	Sig.
		B	Std. error	Beta		
1	(Constant)	3.920	0.142		27.596	0.000
	Factory income	−1.708E–5	0.000	−0.323	−2.058	0.048
	Income from shrimps	3.082E–6	0.000	0.289	1.852	0.074
	Forest area	9.040E–6	0.000	0.254	1.625	0.115

[a] Dependent Variable: Log total losses
[b] $R^2 = 0.54$
Source Household survey

As for the middle income households, forestry and agricultural income increase the vulnerability, and off-farm income decreases the economic losses from disasters, if the damages following the Ketsana typhoon are expected to represent the typical pattern of natural disasters. Income from shrimp farming, and forestry for the rich households are economic activities which increase vulnerability. Also in this case, we find that off-farm economic activities decrease the vulnerability, in this case factory income. Education turned out to be a non-significant factor in explaining disaster losses, regardless of the income group.

3 Inequality: Does It Matter in the Analysis of Vulnerability?

According to Adger (1999), a direct link between inequality and vulnerability shows up in the concentration of access to resources being in a few hands or households, meaning that poorer households have constrained coping strategies. The distribution of resources within a community affects the collective ability to cope with climatic changes or disasters, so more unequal communities could be less capable of addressing natural disasters.

If inequality is an important factor which should be included in vulnerability analysis, two possible approaches could be adopted in our study. First, by determining the degree of inequality comparing the five districts, we have an indication of inequalities of income distribution across districts. Does this information help us to understand the level of disaster losses? Second, can we learn anything about the awareness of climate change's effects among the households, once again by comparing districts?

Comparing the standard deviation to the mean household income in the five districts shows signs of differences among the districts. We find the most equal income distribution in Que Son (a standard deviation of 15 million to the mean of 21 million Dong), whereas Nui Thanh district shows the highest degree of inequality (a standard deviation of 134 million to the mean of 60 million Dong). Table 6 shows the fraction of reported losses compared to the total household income.

As most households (80 %) have experienced substantial losses compared to their income base (above 50 %), and the figures in the other categories are small, there is no statistical significance between districts (Table 6). We interpret this result as an indication that the degree of severity of the losses is not a useful indicator of vulnerability in Quang Nam. Still, we might be able to link the reported income distribution by district to the awareness of climate changes and to the reported compensation of disaster losses.

4 The Institutional Framework: Supporting Disaster Relief Activities?

We now turn to the question of the institutional support. Households were asked whether they have ever heard about climate change, what kind of measures they have taken to reduce potential disaster losses, and whether, in the past, they received disaster relief or not.

In terms of government or NGO support to affected households (Table 7), we can see a significant difference between the administrative support efforts in the five districts. Dai Loc district supported almost all affected households. In Bac Tra My, only 55 % of the households received disaster relief. Table 7 indicates that, in general, despite the official policy of a full coverage disaster relief program, only 88 % of the affected households were compensated by the authorities or by NGOs.

A possible explanation of the picture of variation in the support schemes in five districts is the suggestion that communes are reluctant to compensate high investment losses in plantation forestry and shrimp farming, considered a private business adventures with high risks at stake. Losses of roof or damaged kitchens plus tree destruction are the most common type of losses, and some support for the argument that communes are unlikely to compensate risky investment in plantation activities can be gained by reading Table 8.

In Nui Thanh, where shrimp farming is predominant, eight households did not receive any disaster relief of which two had invested in shrimp farming. Investing in forestry activities is concentrated in Nui Thanh and Bac Tra My, but only in the

Table 6 Loss fraction groups by district

		Loss fraction groups			Total
		0–10 %	Between 10 and 50 %	Above 50 %	
District	Bac Tra My	2	4	26	32
	Dai Loc	0	2	22	24
	Hoi An	3	2	20	25
	Nui Thanh	4	7	37	48
	Que Son	5	2	30	37
Total		14	17	135	166

Source Household survey

Table 7 Disaster relief received or not by district and by loss group ('000 Dong)

Disaster relief received			Disaster losses groups			Total
			0–1,500	1,500–8,000	Above 8,000	
No	District	Bac Tra My	6	2	6	14
		Dai Loc	0	1	0	1
		Hoi An	0	4	4	8
		Nui Thanh	1	4	3	8
		Que Son	3	3	1	7
	Total		10	14	14	38
Yes	District	Bac Tra My	4	8	6	18
		Dai Loc	8	11	4	23
		Hoi An	2	10	5	17
		Nui Thanh	6	10	24	40
		Que Son	8	15	7	30
	Total		28	54	46	128
Total	District	Bac Tra My	10	10	12	32
		Dai Loc	8	12	4	24
		Hoi An	2	14	9	25
		Nui Thanh	7	14	27	48
		Que Son	11	18	8	37
	Total		38	68	60	166

Source Household survey

Table 8 Type of losses by district

		Type of losses					
		Roof, kitchen	Trees (garden, plantation)	Agriculture	Shrimp farming	Others	Total
District	Bac Tra My	8	19	3	0	2	32
	Dai Loc	9	8	5	1	1	24
	Hoi An	19	3	1	2	0	25
	Nui Thanh	16	16	6	7	3	48
	Que Son	15	14	6	0	2	37
Total		67	60	21	10	8	166

Source Household survey

first of these districts have households begun to gain income returns. Here households with substantial plantation holdings have been compensated, regardless of the amount of their investments.

From the household survey, we see that the general compensation was allocated to households primarily in the form of rice, noodles and clothes. An amount of goods equivalent to 1 million Dong was often stated by the households, so major investment losses, due to Ketsana, were never compensated, not even in Nui Thanh.

A more plausible explanation to the variation in support schemes relates to governance or the institutional framework of the districts. When asking households about how they assess local authorities' disaster capacity, we note a significant difference. Approximately 75–80 % of the households in Dai Loc and Que Son were satisfied with the disaster plans drawn up by the communes, whereas only 40–50 % regarded the commune plans adequate in Bac Tra My and Nui Thanh (Table 9). In Vietnam it is not uncommon to regard communities as extension of the Communist party (see "Impediments to Climate-Induced Disaster Management: Evidence from Quang Nam Province, Central Vietnam"), but then to conclude that all villagers are automatically fearful to express their opinions would be erronous. Despite state control, we see people being ready to articulate their dissatisfaction with disaster planning. Referring back to the discussion on income distribution, we note that the commune with the most equal income distribution (Que Son) delivered disaster plans most of the habitants agreed were satisfactory, whereas the picture was exactly the opposite for Nhui Thanh with relative high income inequality and only 50 % of the households supporting the local disaster planning. The association between the two factors might be coincidental.

Answers to other questions concerning people's knowledge of climate change show an equivalent dissymmetry when comparing the answers among the districts. Of course, this could be due to coincidence or to the small survey sample. However, in view of our other questions put to the households, the result simply adds to the overall impression of differences in governance among the districts. The people who are most aware about climate changes are those living in Dai Loc district (Table 10), they are also the ones most satisfied with their district's disaster plans (Table 9), and they were better covered by disaster relief schemes (Table 7). Households living in Que Song knew little about climate changes (Table 10), and the households were not better off than the other districts in relief cover (Table 7). However, they were in general, maybe surprisingly, quite satisfied with the disaster planning in the district (Table 9).

At the other end of the scale, we find Bac Tra My with less that 50 % of the households being satisfied with their district's disaster plans (Table 9) about the same percentage of households claimed that they did not know anything about climate changes (Table 10), and with only 56 % of them having received disaster relief (Table 7). In terms of relative low knowledge about climate changes, households

Table 9 Commune disaster plans according to the households by district

		Does your commune have disaster plans?			Total
		No or not satisfied	Yes	People help each other	
District	Bac Tra My	10	14	8	32
	Dai Loc	4	17	3	24
	Hoi An	8	12	5	25
	Nui Thanh	24	19	5	48
	Que Son	6	30	1	37
Total		52	92	22	166

Source household survey

Table 10 Understanding climate change impacts, answers by district

		Climate change links				Total
		Rising temperatures, more flooding or more often disasters	Deforestation caused by climate change	No answer	Environmental problems, but due to internal factors	
District	Bac Tra My	15	0	17	0	32
	Dai Loc	16	1	7	0	24
	Hoi An	10	0	14	1	25
	Nui Thanh	14	0	30	4	48
	Que Son	9	2	24	2	37
Total		64	3	92	7	166

Source household survey

in Nui Thanh are worse off (more than 60 % of the interviewed households gave no answer to a question asking them, how they would explain climate changes), and they are, as in Bac Tra My, not satisfied with the authorities' disaster planning efforts (only 50 %). But they are better off than the households in Bac Tra My, when it comes to relief coverage (more than 80 % of the households received compensation for disaster losses). The conclusion is then that there is no link between being compensated for incurred losses and appreciating the local disaster plans.

5 Conclusion

We have noted that among the three vulnerability components, income/poverty, inequality and institutional adaptation, the first and the third components showed significant results. House quality is important for poor households, whereas the investment in plantation forestry renders both middle income and rich households vulnerable to natural disasters which might be the outcome of climate changes. We also identify quite significant differences between the five districts where data were collected. If these sample data are to be assumed correct and representative of the situation in the districts, households are better off in Dai Loc than in Nui Thanh and Bac Tra My. The last point is that we saw no correlation between the degree of severity of losses and the geographical location of the households. Of the interviewed households, 80 % of them experienced economic losses greater than 50 % of their income base.

References

Adger N (1999) Social vulnerability to climate change and extremes in coastal Vietnam. World Dev 27:249–269

Canon T (2008) Reducing people's vulnerability to natural hazards. United Nations University, Helsinki Research Paper, No. 2008/34

Chaudhuri S, Jalan J, Suryahadi A (2002) Assessing household vulnerability to poverty from crosssectional data: a methodology and estimates from Indonesia. Columbia University, Columbia Department of Economics Discussion Paper Series 0102-52

Chen TC, Tsay JD, Yen MC, Matsumoto J (2011) Inter-annual variation of the late fall rainfall in central Vietnam. J Clim 25:392–413

Dasgupta A, Baschieri A (2010) Vulnerability to climate change in rural Ghana:mainstreaming climate change in poverty-reduction strategies. J Int Dev 22:803–820

Füssel HM, Klein RJT (2006) Climate change vulnerability assessments: an evolution of conceptual thinking. Clim Change 75:301–329

Hoang LT, Nguyen KV, Mai TT, Phan VT, Pham TL, Le NP, Nguyen XH (2010) Study on variation scope and trends of some basic climatic factors and phenomena in mid-central Vietnam, sub-project 1, project 'Assessing impacts of climate change on physical, environmental conditions and socio-economic development in mid-central Vietnam', Vietnamese- Danish research project, Hanoi

IPCC (2007) Climate Change 2007—Impacts, Adaptation and Vulnerability Contribution of Working Group II to the Fourth Assessment Report of the IPCC. Cambridge, UK

Kelly PM, Adger WN (2000) Theory and practice in assessing vulnerability to climatc changeand facilitating adaptation. Climatic Change 47:325–352

Masozera M, Bailey M, Kerchner C (2007) Distribution of impacts of natural disasters across income groups: a case study of New Orleans. Ecol Econ 63:299–306

Moser C (1998) The asset vulnerability framework: reassessing urban poverty reduction strategies. World Dev 26:1–2

O'Brien K, Eriksen S, Nygaard L, Schjolden A (2008) Why different interpretations of vulnerability matter in climate change discourses. Climate Policy 7:73–88

Republic of Vietnam (2011) Decision of the Prime Minister 9/2011/QD-TTG: promulgating standards of poor households, poor households to apply for stage from 2011 to 2015. http://vanban.chinhphu.vn/portal/page/portal/chinhphu/hethongvanban?class_id=1&_page=37&mode=detail&document_id=98923. Accessed 24 Nov 2012

Van Tan P (2010) Study on developing detailed climate change scenarios for mid-central Vietnam. Sub-project 3, project 'Assessing impacts of climate change on physical, environmental conditions and socio-economic development in mid-central Vietnam', Vietnamese-Danish research project, Hanoi

Rural Households: Socio-Economic Characteristics, Community Organizing and Adaptation Abilities

Ole Bruun

Abstract Based on a comprehensive survey and subsequent fieldwork, this chapter introduces the socio-economic characteristics and common livelihood strategies of rural households in Quang Nam, Central Vietnam. It demonstrates the basic premise of self-reliance in rural society and the decreasing economic dependency on state institutions under the Vietnamese transition to a market society. It discusses present poverty definitions and measures by comparing survey data with the formal economic categorization of rural households. Both the overall characteristics of rural society and qualitative data indicate that the reforms have set in motion a process by which a mix of new opportunities and increasing pressures creates new winners and losers. Second, the chapter draws attention to the nature of interactions between households, local communities and the Vietnamese state. This shows both potentials and limitations of informal organization and community adaptive capacities, and finally, it discusses the issues of vulnerability and adaptation to climatic and environmental change from a household perspective.

Keywords Adaptation capability • Poverty • Formal and informal organization • Stress factors • Livelihoods

1 General Household Characteristics, Ranking and Poverty

Picturesque scenes of peasants wearing dark clothes and large pointed straw hats as they steer ox-driven plows through muddy rice fields still hold true in much of Central Vietnam. Conventional family farming based on paddy rice, vegetables and a little livestock, such as known from large parts of East and Southeast Asia, has since decollectivization again formed the basic livelihood for a broad section of the rural population. Social organization built on the cohabitation unit of the

O. Bruun (✉)
Department of Environmental, Social and Spatial Change,
Roskilde University, Roskilde, Denmark
e-mail: obruun@ruc.dk

O. Bruun and T. Casse (eds.), *On the Frontiers of Climate and Environmental Change*, Environmental Science and Engineering, DOI: 10.1007/978-3-642-35804-3_8,

extended family also tends to be the norm across the region, and public authorities will only calculate economic standing and poverty on a household basis. Yet the reversion to conventional agriculture was a brief interval for many households: the subsequent economic reforms and marketization, inspired by a similar move in China a few years earlier, have steadily impacted on both livelihoods and the structure of production. Under the joint forces of economic reform, industrialization and globalization timeless rural practices are beginning to give way to economic diversification, and rural life is increasing intersecting with migrant work, urban life and migration overseas. Small private landholdings necessitate rural households to either seek income outside farming or to invest in more land for specialized farming. Thus, a rising share of income derives from aquaculture in mainly coastal areas, from plantation forestry in inland areas, or from other specialized cash cropping. Despite industry and tourism now making up three-quarters of GDP, farming continues to play a vital part in economic development as well as in securing basic livelihoods for a broad section of the population.

The material presented here is in part drawn from a comprehensive socio-economic survey, conducted across rural areas of Quang Nam (for methodology see the end of this chapter). The survey was later used as a basis for thematic interviewing among respondents and all relevant local actors, and these again were eventually used as a network of informants for general fieldwork in the rural communities.[1] In this chapter general information and mainly qualitative data are extracted from the survey; for a quantitative analysis see "Livelihood Strategies Under the Constraints of Climate Change Vulnerability in Quang Nam". The survey was conducted across a choice of representative communities in coastal, inland and highland areas, in which the vast majority of households were registered as belonging to the Kinh ethnic majority population, commonly just referred to as 'Vietnamese'. A smaller section of households belonged to ethnic minorities (17 Co/Cua and 1 Co Tu/Katu), given special attention in "Interacting Cultural and Environmental Change: The Co (Cua) Minority of Central Vietnam".

Quang Nam ranks among the middle-to-poor provinces in Vietnam with a Human Development Index (HDI) at 0.56 (2011). In comparison, that of Vietnam in general is 0.59, considerably lower than China (0.69) and only slightly higher than India (0.55).[2] This indicates a very modest quality of life despite considerable gains in recent decades, and a high dependency on both primary occupations and primary social organization. In the survey, a great majority of rural households (73 %, close to provincial average) still depended on agriculture as their main source of income. Another 10 % stated trade, service and tourism as their main occupation, 2 % stated government work, and 12 % were retired. However, using the standard Vietnamese ranking of the rural population into poor (*nghèo*), middle (*trung bình*) and well-off

---

[1] Conducting the survey, interviewing and fieldwork included a total of four months' stay in Central Vietnam.

[2] It has been pointed out that Vietnam's HDI is in fact not impressive (Klump 2004). Vietnam ranked 109 of 177 countries in 2006 and 113 of 187 countries in 2010, with an inequality-adjusted HDI of 0.51, ranking 128 (UNdata).

(*giàu/khá giả*) households, the degree of economic diversification differ between them. Poor households rely almost exclusively on small-scale farming (80 %) or are retired (20 %), while the higher income groups had a substantial share of their income from trade, services, government jobs and other wage labour. In terms of the share of total income derived from agriculture, poor households had 73 %, middle households had 41 % and well-off households had 38 %. Thus, economic diversification is much stronger among the middle and well-off households, or conversely, diversification tends to be a precondition for success.

Vietnam's HDI average of 5.5 years of schooling differs with the survey results. However, a clear correlation between economic ranking and the household head's education was evident. Household heads belonging to 'poor' segment merely had an average 4.3 years of schooling, while 'middle' segment had 6 years and the 'well-off' 7 years. Another clear division, related to the above, is the age composition of the three groups, as poor household heads had an average age of 66 years, compared to 56 years for middle group and 48 years for the well off.

Strong, permanent housing is essential to withstand the extreme weather events such as storms and flooding in Central Vietnam, with concrete houses considered superior to brick houses (less suitable to flooding) and wood-built houses, the latter usually being only simple structures or mere sheds for the poorest population segments. But housing is also a powerful social denominator; all households will build new concrete houses with several floors and elaborate facades when their economic circumstances allow, evident in the high building activity in most districts. Thus well-off households all had concrete houses or, for a small part, brick houses. Middle households had 60 % concrete houses and 35 % brick houses, while poor households only had 15 % concrete houses and 52 % had brick houses. Other material conditions of life showed similar variations, with the distribution of motorbikes between the three groups being 100, 66 and 39 %, again equivalent to the distribution of mobiles phones, being 100, 63 and 31 %. On the other hand a vast majority of all groups had colour TVs, including 77 % of the poor.

In a geographical sense, poverty rates are significantly higher in rural and upland areas than in urban and lowland areas of Vietnam, where the population has benefited disproportionately (Thuc 2007, VASS 2011). This overall pattern is consistent with statistical material from Quang Nam province (see "Paradoxes in Adaptation: Economic Growth and Socio-Economic Differentiation. A Case Study of Mid-Central Vietnam"). In a sociological sense it was evident from both survey and fieldwork that the majority of the desperately poor people in the rural areas (apart from ethnic minorities) belonged to the categories of elderly with few relatives to care for them, households with disabled or chronically ill members unable to work or needing special care, and large households with insufficient labour, often in combination with single-parenthood or disease/disability. A social group that particularly stood out consisted of elderly farmers with very little land, little formal education, smaller households than others (comprising many widowed elderly and elderly couples living separate from the younger generation), and significantly poorer housing (houses of brick, clay or wood), making them both economically and materially vulnerable.

Inevitably, there are political implications of economic ranking. The Vietnamese local governments will only categorize those households suffering from abnormal conditions as 'poor' (thus meriting minimum public support), while normal, able-bodied people of working age are supposed to care for themselves at any time. In the absence of a rights-based approach, such as rights to land, livelihoods or food security, rural people do not perceive of state support as other than situational endowments, such as in emergencies after disasters (see "Impediments to Climate-Induced Disaster Management: Evidence from Quang Nam Province, Central Vietnam"). Cultural and historical circumstances dictate self-reliance and the long interval of communist rule did not change these basic premises. In accordance with the policy above, the Vietnamese poverty line (from the Government Statistical Office, GSO) is very low – at the time of the survey 0.33 USD per day (210.000 VD = 10 USD per month) as compared to the UN standard of 1.25 USD—thus signifying extreme/food poverty. Given the current prices of rice (8–13,000 VD/kg) this would ensure nothing more than a daily intake of calories, equivalent to 0.65 kg of medium-grade rice or 2,300 calories per day.[3] To bridge the gap between a food poverty line and real-life conditions local governments may use slightly more realistic values when determining who is eligible for public support. Some used a 400,000 VD line (e.g. Dai Loc) and others refused to calculate income and instead evaluated the quality and equipment of the family house by personal inspection (e.g. Que Son). Certain groups are entitled to support. Disabled war veterans for example receive a small monthly allowance from central funds, and poor elderly people may receive a small monthly allowance of a yearly gift of a little cash, rice and noodles from local government. Housing programs are the most common, and most often the only, poverty alleviation activities of local government, mostly consisting of low interest loans, and sometimes a monetary contribution, for house building. Previously, poor landless households might be allocated land by their commune, but after the 'privatization' and land rush in recent years this has stopped.

An observation from the survey was that the formal ranking into the three income groups above did not correspond to self-evaluation.[4] Deviation was strongest in relation to the category of 'poor': no less than 51 % of rural households perceived of themselves as 'poor' as compared to a provincial average rate of 15 %. The average total household income from survey was 34.8 million VD, corresponding to only 1.1 USD per capita per day, and simple calculation of self-reported income would place an astonishing 80 % of households (132 out of 166) below the

[3] The Vietnamese food poverty line uses a standard of 2100 calories per day. Poverty levels have been modified regularly by the Vietnamese government and in April 2011 it was raised to 400,000 VD per month, corresponding to 0.63 USD per day. This new poverty line produced a national poverty rate of 12%. Applying the 1.25 USD level produced a poverty head count of 16.85% in 2008 (WB 2010).

[4] Yet, as pointed out by local governments all data on household income are incomplete and subject to conservative assessments by the households themselves; this also showed in many villagers' perception of climate-change adaptation research as related to poverty alleviation efforts and government aid.

international standard. It should be reiterated, however, that the total yearly incomes as stated by households are subject to considerable uncertainty. Many key data are well known by local authorities and reported to central level in yearly statistics, but other forms of income such as relatives' remittances and business proceeds are only reluctantly reported. Vietnamese authorities, for their part, have assimilated the Chinese practices by calculating migrant work by unmarried household members as regular household income, despite the migrants living away from home, thus impacting significantly on official poverty levels in rural areas.[5] Whether this practice is reasonable or a distortion of facts depends on a number of domestic factors, but in any case the migrants will need part of their income to sustain their own lives.

As an alternative perspective on local poverty levels, a qualitative approach was used. A range of local people in diverse localities were asked to assess a minimum income to meet the most basic needs[6]; depending on geographical area this leaned towards 0.5–1 million VD (24–48 USD) per capita per month in 2011, lending some credibility to the perceived poverty rate above. Survey and fieldwork were conducted during a period with increasing macro-economic instability in Vietnam and increasing inflation, reaching 10 % in 2010 and peaking at 23 % in the autumn of 2011. Particularly food price hikes inevitably encroached upon poverty alleviation efforts and food security across Vietnam (Coxhead and Vu 2012), and coinciding with the aftermath of the disaster from Ketsana in 2009, many poorer households were subjected to multiple shocks.

Those households already belonging to the middle or higher income groups proved far more capable of taking part in the overall economic growth in Vietnam. Again, based on self-evaluation, the high-income group scored 1.54 and the middle group scored 1.14 on a scale of increasing or decreasing household living standards (The overall score was 0.60, indicating general experience of gradually rising living standards).[7] In contrast, the lower income groups had higher frequencies of households having experienced stagnant or decreasing incomes (some of these could of course have fallen from higher income groups), and the poorest section scored as low as ÷0.05. Geographically variation was evident from the coastal district of Hoi An (scoring 0.84) to the coastal/upland district Nui Thanh (0.67), to the lowland district Que Son (0.56), and to the inland districts Bac Tra My (0.44) and Dai Loc (0.33). These latter two districts contain many poverty-stricken communes with official poverty rates as high as 30 %, emphasizing persistent territorial inequalities.

Place matters, but as seen in the survey it is not in itself a consistent factor: not all vulnerable people live in the most vulnerable places, while even fragile

[5] Local governments may use a minimum income of 1 million VD per month or an estimate for calculation of total household income. Even when marrying, migrants' income will be included; only when migrants achieve local registration in their new place of residence will they be regarded as a separate household.

[6] Those expenses stated were primarily for food, shelter, health and school fees.

[7] A scale indicating self-evaluation of change in income (increase much: 2, increase little: 1, stable: 0, Decrease little: ÷1, decrease much: ÷2)

physical environments may provide access to wealth for some people (Chaudhuri et al. 2002; O'Brien and Leichenko 2000, p. 224). Despite coastal districts having higher general expectations of rising incomes, the household's present social status appear to be by far the most significant factor. Differentiated experiences of participating in economic development call into question the present state poverty alleviation, or 'pro-poor growth' in Central Vietnam. A concoction of high inflation, economic instability, rising expenses for education and health care, and 'loose policies' usually hits hard on the poor (e.g. Singh et. al. 2003, p. 28; Kang and Imai 2010; Herrador 2012). Particularly potent in the present context is the high rate of rural population living far below, just below or close to the perceived poverty line; at the same time this segment presumably includes the most climate-vulnerable households.

2 Composition of Livelihoods

With the inception of Doi Moi policy in 1986 a range of new economic opportunities began to emerge. At the same time marketization encroached on small farming incomes. For the poorer population segments, the conventional products of paddy rice, vegetables and a few chickens or ducks, and in some cases a pig, provide the basic subsistence and a few items to trade for cash, but not on a scale to lift the household out of a meager existence. Despite improvements in rice yields (see "Is Climate Change a Reality for Agriculture in Quang Nam Province?") and increasing emphasis on specialized cash crops such as maize, melons, papaya, beans, cashew and cinnamon, the small plots of average households (approximately 600 sq.m.) will set the limit. Thus, for small farmers, improvements in agricultural production may only keep pace with rising living expenses, including rising fees for health and schooling. Strategies for achieving a higher living standard must embrace either farm expansion for cash crops, wage labour, or investment in business. The occupational data above merely indicate what was stated as the main household occupation. In actual practice, households may utilize a broad mixture of local resources and income sources, for which 'livelihoods' (e.g. Chambers 1995; Wisner et.al. 2004) may be a more appropriate term than income level.[8]

The majority of the land-owning rural population will seek wage labour income to supplement conventional farming (which may be run as usual or entirely oriented towards home consumption). One or several household members may engage in migrant labour or pick up whatever local wage labour opportunities are available. Young people's migration to coastal areas has become the norm; their work input may not be decisive at home and after leaving school there is little to

[8] Livelihoods may refer to the command over incomes and bundles of resource that may be used to satisfy needs (e.g. Wisner et. al. 2004). For Robert Chambers (1995) 'livelihoods' are distinguished from poverty by including other aspects of deprivation, such as vulnerability, seasonality, powerlessness and humiliation.

keep them occupied in the villages. From the age of 20, many leave for further education or migrant work depending on family assets, and for the age group of 20–40 nearly half of them will have left their villages in most areas; in the poorer districts as much as 60–70 % of the young will leave. Some go to work in the tourist industry, others find construction work on the coast from Danang to Tam Ky, and yet others are employed in the textile industry around Saigon.

Many kinds of day-labour are available in the rural areas of Quang Nam. Agricultural day labour is now very common as the larger farms depend on it at various stages of production. The rising plantation forestry in particular requires labour for planting, weeding and harvesting fruits, rubber, coffee and acacia trunks, but also aquaculture increasingly depends on hired labour. Other common jobs are house-building, painting, carpentry and casual jobs around the better-off households' fields and property. Some small-scale rural industries employ locals on a daily basis, but they are more prominent in coastal districts. In most places, the district governments employ local labour, such as for cleaning and cooking in government buildings and for maintaining simple dirt roads, small bridges and river passes. They typically pay 30,000 VD for women and 50,000 VD for men per day (2010/2011). Payment for day labour is relatively low throughout, ranging between 30,000 VD and 80,000 VD in rural areas and up to 120,000 VD per day in construction work on the coast. Moreover, day labour opportunities are unevenly distributed over the year and generally far less available than desired. Although many men and women in principle are available every day, they may only work a few days every month, providing only a very modest income and causing seasonal deprivation.

The poorer population segments may utilize a range of marginal resources. They may gather crabs or tiny water snails in lakes and water courses to sell for a little cash in the market, or they may collect firewood to bring to people's homes. In mountainous areas NTFPs may be available, but due to the retreating natural forest they are only available to the outlying communities. The grazing of domestic animals in plantation forests and unoccupied land, which provides opportunities for both poor and better-off households, is of rapidly increasing economic importance across the province.

As suggested by self-evaluation, qualitative fieldwork also pointed to an accelerating, or self-strengthening, differentiation process. Households rising above the basic livelihoods of small-scale farming have a range of opportunities for investment and capital accumulation. Common strategies include buying or contracting land, reclaiming land for plantation forestry (see "The Push for Plantations: Drivers, Rationales and Social Vulnerability in Quảng Nam Province, Vietnam" and "Impediments to Climate-Induced Disaster Management: Evidence from Quang Nam Province, Central Vietnam"), or investing in shops, restaurants and other businesses. A widely used strategy is to use surplus funds, such as from migrant labour, for investments in small household businesses, typically shops, restaurants and repair shops in villages or along main roads. Incomes from small businesses may not be great, but may provide a welcome cash supplement to small farming. The most substantial household investments are placed in shrimp and crab farming in coastal areas and in plantation forest land in central and mountain districts: they all benefit from the low labour costs and lack of regulation. Although inherently risky investments in a hazard-prone environment, if successful over just a short time span, they may bring very large profits, enabling households to abandon conventional farming altogether. The tourist industry is yet another sector of rising importance, particularly in coastal areas, where rural households invest in shops, restaurants and service businesses.

On a provincial basis the contribution of agriculture to GDP is decreasing as a manifestation of industrialization and general economic growth. Yet, when participants in the survey were asked about their continued dependency on agriculture a somewhat surprising pattern emerged: 37 % stated increasing dependency as against 29 % stating decreasing dependency over the last 10 years. Another aspect of industrialization, which may be relevant for interpreting the above data, is the relative aging of the farming population. Many households now consist of pensioners, who keep farming on a very modest scale, while the younger generation has left the rural community.

In any agricultural community land tenure is a crucial issue, wrought with tensions, inequalities, and forms of domination. At the outset of the Doi Moi policy, land was collectively owned and then allocated to private households according to a mixture of pre-war ownership and present household size and composition. Since the Land Law of 1993/2004, enabling the selling and transfer of land as well as the reclamation of ‘unused land’, a new segment of large scale farmers has emerged which is on the rise. These have several thousand m^2 of farm land and up to 10–15 ha of forest land. It is presumably this new trend in agriculture that strikes through in poorer and wealthier rural households and where there is a stronger connection to agriculture than middle households.

Until a few decades ago, everyone shared poverty in rural areas. Today, differentiation is evident: villages consist of both utterly deprived households living in shacks and wealthy household in lavish villas with every possible comfort, including cars, and sending their children to study in Australia, USA or Europe. In the survey this showed, for instance, in a marked correspondence between economic ranking and land holdings. The well-off households on average had 2.5 times larger land holdings than middle households and 5 times larger land holdings than poor households (taken as groups despite not all being farmers). This was equally evident in the stated yearly total incomes for the three groups, which amounted to approximately 51, 27, and 12 million VD. A small group of households in each locality nevertheless appeared to have amassed land on an unprecedented scale, boasting land holdings of up to 60 ha. This new rural elite, consisting of former or present officials, well-connected individuals, incoming investors, or highly entrepreneurial individuals allegedly using bribery of commune administration to get access to ‘unused land’ (see “Interacting Cultural and Environmental Change: The Co (Cua) Minority of Central Vietnam”), has become a main factor in the rural day labour market. That corruption has become a consistent problem in land management is documented in international reports (e.g. WB 2011; Brown 2012) and widely reported by rural people across Quang Nam during fieldwork, particularly concerning access to ‘unused land’

and payment for land registration (Bruun 2012). Other instances related to outright land grabs where, for instance, a surveyed household with attractive beach land had its land reclaimed by local government, only to be sold to a holiday resort investor, and offered compensation at a fraction of its market value.[9]

3 Household, 'Community' and State

The majority of households had lived in their present homesteads for the last 20 years or more, indicative of a high degree of continuity in the rural communities. Despite all having experienced severe weather and floods over the last several years, very few households reported to have left their former homesteads for such reasons; a few cases related to landslides or constant flooding of houses. The common life cycle events of births, marriages, retirements, deaths and land inheritances were the main factors for change. Overall, only a small portion of households had moved for other reasons, of which the most common factors were the opportunities to get better land, improve the economic situation, environmental interventions and land reallocations by government. Several thousand households had been affected by the construction of the huge Phu Ninh Lake and hydropower station near Tam Ky up to its completion in 1990 (see "Natural Resource Management Impact on Vulnerability in Relation to Climate Change: A Case in a Micro-Scale Vietnamese Context") where most were moved to inland districts.

Thus the majority of Kinh communities were relatively stable, with most families tilling the same land as their ancestors and each village consisting of interlocking webs of kinship relations. This is not a region characterized by the 'closed corporate peasant community model' (e.g. Rambo 2005, p. 118), in which surname groups tend to dominate entire villages such as in parts of north Vietnam and China. Yet fairly large groups of people being closely or distantly related and sharing the same family name may be found in most villages, due to the predominantly patrilocal, and certainly also patriarchal, form of organization. Smaller lineage-based hamlets are found, some tracing their ancestors back several hundred years and still entering new lineage members in their precious ancestry books at yearly ceremonies.

Ancestor halls, most often the size of family houses and with similar gardens or enclosures around, are found everywhere. They are richly decorated in traditional designs and symbols, and frequently with inscriptions in Chinese characters. They bear witness to traditional rural organization as well as to a cautious opening towards greater religious freedom in Vietnam. The rising number of newly-restored or rebuilt ancestor halls in Central Vietnamese villages also testifies to the hugely important economic and social exchange between locals and wealthy migrants

[9] The household was offered a compensation of 50,000 VD per sq. m. as opposed to an estimated market value of several million VD.

overseas, who regularly will travel back to their ancestral lands. Some halls constructed in the prewar period are of immense proportions and appear like regular Buddhist temples with huge walled courtyards and lush gardens. Other halls are distinguished by their exquisite interior decoration in marble and the finest hardwood, comparable to the best temples in Asia. Described below, the communal significance of these ancestor halls is less lucid.

A characterization of present rural administrative units, including villages and hamlets, as 'communities' may obscure the fact that the Communist Party organization is still firmly in place and dominates all aspects of rural organization. The formal organization stretches downward from the Provincial People's Committee to District and Commune Committees. Communes, villages and hamlets are again reminiscent of production brigades and teams, transferred to a civil administration, but still under exclusive Party command. In fact, the generic term 'community', so common in international literature, is not used at all: both local governments and people will instead use the concrete term for the unit in question, such as 'village', 'commune' or 'district'.

In the ordinary Kinh communes state and party organizing still dominates all aspects of life. An impressive range or pro-government mass organizations are present, mostly under the umbrella organization, the Vietnamese Fatherland Front. They include the Communist Party; Ho Chi Minh Communist Youth League; Peasants', Women's, Elderly's, Workers' and Veterans' Associations; Classmate and Age-group associations; Home Guard; and more. However, none of them could gather more than a small fraction of households in any village. The only exception was the Women's Association, which might gather a fair number of women around its activities, including various training courses and small-scale business initiatives. Considering the highly patriarchal organization in households and to a great extent in society at large, the Women Association continues to play an important role, such as to mobilize women against what was repeatedly stated as excessive drinking and gambling among their husbands.

Otherwise, formal organizing is weak, while at the same time non-state actors are few and far between: 'civil society' development remains a sensitive issue.[10] Neither traditional institutions such as Buddhist temples and lineage organization, nor new civil society organizations have great rooms of maneuver. A range of international NGOs are present in Quang Nam and Central Vietnam, and their numbers have been growing. The most prominent are Red Cross, Care, World Vision, Ford Foundation, EMV, WWF, IFID, Malteser, and DED, all participating in cooperative organs. Buddhist charity organizations, mainly the Vietnamese Buddhist Association, appear active on a general scale in the province, particularly in terms of disaster relief. Most NGOs work with education, poverty alleviation, small-scale construction and other projects in primarily the poorer districts, typically in upland

[10] According to the UNDP, the less conducive factors for civil society development in Vietnam are limited political competition, high levels of corruption, limited basic civil rights and press freedoms, and restricted legal environment for civil society (UNDP 2011, p. 46).

areas. They work closely with public authorities and in accordance with government programs as is the norm in Vietnam. In many cases these become extensions of government, such as when running public schools and kindergartens. They are still seen to provide powerful pro-poor models and to show altruistic personal conduct as conscious alternatives to the 'market-driven' practices of both public and private institutions.

Apart from the activities of externally funded and controlled NGOs, local and spontaneous civil society organizing is limited, in fact virtually absent. Pre-Communist forms of cooperation were found, such as among families belonging to a given ancestor hall. As elsewhere, people who have once shared a household remain close and exchange funds, goods and labour: obvious examples are a son helping his elderly parents with a new house or siblings helping each other out. The broader lineage group may be important for exchanging labour, and in case of disasters the better-off may provide some assistance for poorer members of the lineage. Kinship organizations are controversial in any socialist setting, however, and the true extent of kinship favours and obligations may not be disclosed, particularly when they involve people of formal authority. The closest cooperation was found among some tightly knit single-surname hamlets that exist in parts of Quang Nam.

Non-kinship based informal cooperation can take many forms depending on income levels and specific livelihoods in local areas. The booming shrimp-farming in coastal and riverine areas has encouraged many cooperative projects, where small groups of farmers pool their land and dig out basins to raise shrimps; profits are huge but so are risks, and not all succeed, such as when the river water quality proves inadequate and the shrimps will not grow. Similar constructions may be used for investments in small fishing boats, since fishermen now need to go further out for a good catch (see below). Particularly among the poorer population segments rotating credit schemes may be found, for instance enabling participants to have a sum of 1 million. VD at one point during a ten-month cycle of 100.000 VD payments, which provides investments for small businesses and shops. Overall, however, these initiatives appear too few to be significant and rarely enjoy official support. Although a range of formal credit institutions are operating in rural areas (Takashi 2009), they were mostly seen to support agricultural production, including grain production and livestock, or house building.

4 Organizing for Climate and Environmental Change

A fundamental question is whether the existing system of territorial and political organization in Vietnam is adequate for future adaptation to climate and environmental change. The formal structure derives from the communist organization of society; rural communes have broken up as farming units, however, and their inhabitants are increasingly transferring to non-agricultural activities. Organizing remains the exclusive privilege of state and party authorities, and conventional or instinctive forms of organizing, such as those relating to religion, kinship and profession, are kept under control. Hence, it may with good reason be asked if such top-down approaches provide sufficient space for local-level adaptation efforts—that can enable the contribution from civil society and participatory approaches—or if the present political apparatus is really out-maneuvering any spontaneous organizing.

To judge the performance of existing government and political mass organizations in preparing local communities for climate change and natural disasters, the survey inquired about people's participation in their meetings and perceptions of their activities. On the question if climate and natural disasters had been discussed on such meetings, answers indicated that they were mostly used by arms of government, either commune authorities or district disaster prevention teams, to pass down information. There was a general perception of good advice on a range of preventive measures, either before or during disasters. People were advised to strengthen their houses, secure their belongings, store food and water in safe places, not to send their children to school during storms, and prepare for moving away from flooded areas if necessary. Women in particular were encouraged by the Women's Association to take action on behalf of their families. Health and hygiene during floods were also commonly addressed, as latrines may overflow and cause severe health risks. There was special advice for occupational groups, such as for farmers to adapt their farming calendar and crop cycles to local flooding patterns, to improve irrigation systems in order to secure a crop in the dry season, and to make safe places for domestic animals. Shrimp breeders were advised to protect their stock against possible intrusion of salt water, and fishermen were advised not to go to sea before storms, or if possible, seek casual labour during the entire typhoon season. Some communities in addition were advised to plant trees to provide shelter for storms. Overall, because of available information and not least to a vast historical experience, there was a large body of local knowledge on preventive measures. The extent and timeliness of local warnings was sometimes questioned, in line with similar findings from another survey (EMW 2009, p. 22). Particularly in areas where hydropower construction adds to existing flooding risks, there was frustration with the lack of communication. In actual practice, however, people tend to rely mostly on weather forecasts on TV combined with information from village loudspeakers (though not all hamlets have them), until that critical moment when power goes off and people are left on their own.

Both the survey and fieldwork showed that within hamlets and villages, people commonly help each other during disasters, and that most of the wealthier households as a matter of course provide room for those people in their immediate neighborhoods with inadequate or flooded houses. Relations are cordial and people quite freely associate between households, coming and going without much formality. Very poor elderly or disabled people may also receive a little help from the households around them. But apart from such common sociability there is not a strong spirit of cooperation beyond the immediate family and kinship relations.

Answers were hazy on the question of whether local people and their organizations were invited to discuss responses to natural disasters and climate change.

Less than a third answered positively and few details were given. Hardly anyone responded positively to the question, if suggestions from local community members were transferred to higher level authorities. Similarly, no answers were forthcoming when asked if there were any other way people could offer suggestions, or if local organizations generally received feedback on their suggestions. Both the survey and additional interviewing asserted that there is no encouragement for local people to organize spontaneously in order to carry out adaptation measures such as dikes, embankments, protective forest belts, joint house-building, and the like. Up until now, all climate change adaptation and related construction activities have exclusively been questions of government planning—and eventually matters of government funding—while at the same time local governments most often stated that they lack the funding required. Moreover, local administrations at district and commune levels generally claimed not to have the authority to independently carry out adaptation projects.

Hence, when typhoons, floods, or other natural hazards strike and people are in great distress, they still tend to look towards government, since there is no alternative. Individual villagers, especially poorer people, may expect some form of compensation for their losses. After de-collectivization, however, commune administrations decline any responsibility; some expressed that 'people still expect us to do everything—they must learn to take evasive action themselves and to build stronger houses'. After great disasters, such as the Ketsana typhoon in 2009, large amounts of charity and emergency aid have flown into Central Vietnam. The survey showed that local commune and district governments were by far the most important sources of emergency aid, including rice, noodles, drinking water and a little cash. Although aid came from a variety of sources, however, nearly all aid was distributed by local governments, since they maintain a monopoly on allocating aid based on the assumption that they know best concerning the needs of households. Thus, the end receivers of aid have no knowledge of the original source. Only the Buddhist Association was allowed to distribute charity aid independently. Irregularities in the distribution of aid and accusations of misappropriation flourished all over, though in certain districts more than others.

Considerable differences were found in people's perceptions of their local governments, particularly in relation to disaster relief, level of information and general trust. Again, these factors call into question the overall efficiency of government efforts, such as in adaptation. After a long period of centralized government, which has marginalized traditional forms of power and authority in society and organized all local communities into standardized and manageable units, any expectations for problem-solving, whether of an economic, social or environmental nature, point mostly towards government. Today, when public authorities retreat from certain areas of responsibility due to the market reforms, there is a vacuum of local institutions to meet environmental and climatic challenges (a common problem as pointed out by, e.g., Adger et al. 2009). Hence, in the present setup, local adaptation tends to be either stalled or entirely left to market forces.

5 Present Stress Factors and Responses

The survey indicated a variety of responses to climatic and environmental stresses; some were quite conventional while others were unexpected. First of all, and as anticipated, respondents seldom related to climate change in the abstract sense, but instead paid attention to separate occurrences like typhoons, heavy rains, flooding, droughts, seasonality and so forth. Most households could remember the events of severe typhoons or floods going back several decades. But although the concept of climate change and the commonly suggested effects of such were known to some villagers due mainly to TV, it seemed too abstract for others.

A fundamental presumption is that the rural population in Central Vietnam is subjected to a range of stresses, the individual contribution of which is less important than their total impact; conversely, improved livelihoods in general will enable people to be better at facing any individual stress factor (see "Climate Change, Adaptation and the Environment in Central Vietnam"). Overall, and not surprisingly, there is a general perception of land scarcity: less than a third of households regarded their land as sufficient, the rest being equally divided between finding it 'not sufficient' and 'very insufficient'. When asked about what households saw as the main obstacles to generating income in their communities, clear differences appeared between higher and lower income groups. In accordance with the above presumption, the segment of poor farmers showed a fairly equal ranking of a range of stress factors (notably land, water, capital and employment insufficiency, disease, climate/weather events and 'others'). Presumably, this indicates that this particular population segment faces a greater range of stresses that impact their daily lives than the other groups. However, after disasters the lack of capital was particularly destructive for the poorer population segments. In reality this meant a lack of funds for seeds, fertilizer, pesticides and basic means of production. In these cases, material poverty and poor housing were most often equivalent to households losing everything. In contrast, the well-off households had a much higher ranking of land and water scarcity, followed by stresses like pests, market conditions and lack of capital. Middle households placed themselves between these extremes.

The survey itself was conducted in the aftermath of the Ketsana typhoon which was the worst in many years. Perhaps this contributed to the fact that climate/extreme weather events scored higher than other livelihood stresses (see "Health Impacts of Climate and Environmental Change: Awareness and Challenges to Adaptation", Table 8). Yet interviewing indicated that all segments of the rural population are faced with a concoction of problems and risks, of which extreme weather events is neither a new, nor presently a single dominant factor in rural communities. Among the other common stress factors in rural communities, a lack of cash income opportunities was particularly addressed by the poorer segments (apart from the group of elderly). Disease and disability also scored quite high (see "Health Impacts of Climate and Environmental Change: Awareness and Challenges to Adaptation"), perhaps in part related to the aging farming population. Other common stress factors, notably policy and market constraints and pests scored equally, but lower than those above, although with some variation between income groups.

Apart from these obvious stress factors, a range of other factors were mentioned as causing distress for families or encroaching upon their budgets, such as poor roads and lacking means of transportation in some areas, particularly in relation to acacia forest plantation, which demands access for trucks (cxamples are given in "Natural Resource Management Impact on Vulnerability in Relation to Climate Change: A Case in a Micro-Scale Vietnamese Context"). Many households were burdened by disabled family members needing constant care. Other stresses related to rising health fees and rising school fees were those that particularly burdened families with many children such as illicit fees to public authorities. In particular, these two areas of public goods have exploded and caused alarm on a national scale (Brown 2012). Insufficient education, lack of proper equipment, instable power supply in many localities, and disease in domestic animals were mentioned by others.

Coastal areas are sensitive to storms, flooding, and an intrusion of salty water, that carry the potential of destroying dikes and crops. The growing aquaculture in these places has provided large incomes, but it carries greater risks of losing an entire production of shrimps or fish in storms. A very prominent problem reported in the coastal areas around Hoian was the pollution of rivers and streams that destroys the lucrative shrimp farming in parts of the river delta and wetlands. This was caused by hotels in the growing tourist industry emitting unfiltered water, but chemical and petroleum pollution were allegedly contributing as well. Fishing villages along the coast were further affected by a gradual depletion of fish in coastal waters (reportedly related to overfishing rather than climate change). This demands that fishermen go further out and stay away longer for a decent catch, adding to the risk of fishing. Particularly the poorer segments in these villages, fishing from traditional 'baskets-boats' (merely large weaved baskets, shaped like cooking pots, approximately 6 feet across and water proofed with natural rubber) were finding it increasingly difficult to make a living from fishing. Yet these communities were better off in terms of the other income opportunities offered in the coastal region.

Many mountain communes had suffered from flash floods, landslides, loss of farmland and typhoon damage in several successive years, which local governments tended to interpret as climate change upsetting poverty alleviation efforts. The sequence of events is very difficult to establish, however, since a range of environmental interventions are constantly changing the landscape. Deforestation in upland regions will potentially affect seasonal flooding patterns, soil erosion and conditions for new plantation forest. For instance, in the mountain areas of Bac Tra My, extensive forest felling during several decades and the replacement of diverse forest with mono-culture production forest might be partly responsible for the increasing occurrences of flash floods, which regularly destroy fields, dikes and crops. New farm-based forest lots were planted with acacia trees for paper production, and cut every four-to-six years to ensure the families a steady income. The result has been young and vulnerable forest growth which is easily swept away when the stronger typhoons hit, as was the case in 2009.

Hydro-power construction on major upland rivers is also emerging as a crucial issue in Quang Nam. An insatiable demand for electrical power in Vietnam has accentuated the hydro-power potential of Quang Nam province. Quang Nam province has approved some 58 hydro-power projects (LAV 2009), of which 10 are large units on the major Vu Gia and Thu Bon rivers. This will inevitably affect the seasonal flow in major waterways. Interviews indicated that many villages had experienced rising water levels during flooding after the completion of hydro-power stations on their section of the river. For instance, in Dai Loc a recent hydro-power project reportedly had caused the water level to rise an extra meter during the seasonal flooding. This meant that traditional flooding measures in river bank communities were insufficient, and heavy sedimentation of the annually flooded farmland was induced further.

Substantial economic losses due to disasters were sustained in those districts covered by the survey (see "Livelihood Strategies Under the Constraints of Climate Change Vulnerability in Quang Nam") and most households were affected, although in diverse ways. A main distinguishing factor for personal safety and comfort is the quality and size of the family house. The larger concrete houses of the better off families can better withstand storms and water masses, as well as allow space for an elevated platform under the roof, to which the family can retreat with their most valuable property during flooding. The lower and weaker wooden houses of poor families, however, are often just shacks built from second hand materials and with thin steel plated roofs. These are more prone to be entirely flooded and the family will inevitably have to escape to higher ground or seek shelter with more fortunate neighbours.

When disaster strikes the poorer households are, in a material sense, much worse hit than others. The value of their losses may be quite small, but it may still affect them proportionally much harder, such as when the steel roof is blown off their house or when they watch all their belongings being carried away by the water masses: crops, domestic animals, farming equipment and personal property. As uttered by a poor farmer, 'of course it is much easier for the wealthier people in natural disasters. They have good strong houses and can always make it; people like us are just blown away and everything we own is destroyed'.

If climate change triggers a higher frequency of extreme weather, however, everyone suffers. When traditional flooding measures such as elevated storage facilities and platforms for domestic animals are no longer enough, or when typhoons hit with 'higher-than-usual' power and frequency, even those with the means to protect themselves will be affected. In a financial sense, wealthier households obviously tend to be worse hit by natural disasters. Having more to lose, the value of their losses is far bigger: they have more land, more valuable crops consisting mostly of cash crops, greater value placed in domestic animals, and far better equipment than the simple tools of small farmers. Yet their economic losses tend to impact through the local economy down to poor farmers and landless labourers. Small rural enterprises, forestry plantations and aquaculture farms, that all hire local hands, may also be put out of business temporarily or permanently to the detriment of local wage labour opportunities. Many poor households rely heavily on this cash income and they tend not to have any reserves to rely on if they temporarily disappear. Thus, already living at the margins, they may suffer long-term setbacks in their livelihoods when, for instance, they lack cash to rebuild homes or to buy new essential farm tools, seeds, fertilizer and pesticides.

Experience from around the world shows that natural disasters may also provide new opportunities for some, such as by land grabs or a simple reshuffling of resources. During floods villagers may seize what will be stranded around their property, including valuable hardwood timber from illegal logging, boats, animals and objects of all kinds. Uprooted trees flowing downstream might give opportunities to marginal population segments, such as to poor day labourers, who may cut them up for sale as firewood. Other perverted forms of profiteering from disasters were referred to by locals as, 'during disasters some people get rich', since a range of economic and material opportunities follow in the wake of disaster. A vast influx of international, state and private assistance, including huge amounts of cash money to be allocated or spent by consecutive levels of government, certainly provides opportunities for a privileged few in local society. Although merely anecdotal evidence may be provided for these abuses of power and funding, the survey certainly indicated a meager value of disaster relief distributed to poor families.

6 Combined Stresses and Vulnerability

It remains to sum up the various data and observations in order to assess qualitatively how combined stresses strike through as vulnerability to climate and environmental change among certain groups. The heart of this matter is obviously the impact of climate disruptions on people and communities, particularly in rural areas, since climate change is commonly seen as endangering poverty reduction (e.g. Sperling et al. 2002; Oxfam 2008; McElwee 2010, p. 16). As noted in "Climate Change, Adaptation and the Environment in Central Vietnam", however, from a local community and environmental justice perspective, we see no point in a categorical distinction being made between climate change and environmental interventions.

Again, picking up from "Climate Change, Adaptation and the Environment in Central Vietnam", the social vulnerability perspective emphasizes a distinction between natural hazards and people's vulnerability to hazards (Wisner et. al. 2004). This points to social structures and the overall political-economic organization of society as the causes of vulnerability—as much as they facilitate responses to external conditions (e.g. Adger 1998; Gaillard et. al. 2007; Schipper 2007). It has further been shown how the Vietnamese 'hydraulic' society historically has engaged the environment in transformational schemes, continuing today in massive hydropower constructions and dam and dike building across Central Vietnam: combined with the forces of marketization reforms and globalization, in the present rural society few livelihood conditions or resource entitlements are spared from change. Hence, defining social vulnerability as primarily a characteristic of society and its underlying social, economic and political conditions, and striking through as a resulting, inadequate capacity of individuals or groups to cope with and recover from the impact of hazards, is indeed justified by the multiple, dynamic pressures that each and every household is subjected to in Vietnam's present phase of development. A complex mix of interrelated factors generates vulnerability, essentially as a 'lack of mutuality' between environment and human activity over time (Bankoff 2003). Processes of differentiation and marginalization result from political neglect, poor regulation, inadequate redistribution or loose government (Eriksen et. al. 2011; Bruun 2012; Marino and Ribot 2012).

Poverty and vulnerability tend to occur simultaneously. Yet, as often stated, although poor people are usually among the most vulnerable, not all vulnerable people are poor (Moser 1998, p. 3). It was shown above that apart from the decisively poor, a high rate of rural households live just below or close to the perceived poverty line, which in actual practice broadens potential vulnerability. Vulnerability may extend across multiple dimensions like livelihoods, household composition, ethnicity, geographical area and a range of other features, and no single set of criteria will cover them all. Applying those insights above and drawing on first-hand experiences across Quang Nam, we may qualitatively identify a range of social-group and adverse-condition combinations.

(a) The group of elderly farmers identified above as among the poorest, also stand out as among the most vulnerable to hazards. With little land, deficient labour, few assets, and often sub-standard housing they may suffer damages beyond their coping capacity.

(b) A large group of households are essentially 'unfit for fight' in a rough market society because of a range of stresses relating to disease, disabilities, alcoholism, social problems, too many children, single parenthood and so forth, and they tend to fall through the wide mesh of official poverty definitions. Already subjected to both internal and external pressures, further stress from hazards may exclude them from participating in the general economic development, or worse. Unfortunately, there is a great deal of these households, since the above conditions are universal companions of social life. Both the survey and fieldwork indicated that as much as 20–30 % of rural households may in fact suffer one or more of the adverse conditions.

(c) Many households are subjected to adverse or changing physical conditions in their native place. These can relate to a broad category of increasing occurrences of extreme weather events, landslides, flooding, deforestation, hydropower construction, population movements, river bank erosion, etc. Whether due to climate change, man-made environmental change or a combination of each however, the losses to such hazards are generally not compensated for. For instance, households losing farmland due to sedimentation, landslides or river bank erosion may no longer be allocated compensatory land, because land itself has been subjected to marketization, while environmental management tends to lose out to economic exploitation. Hence, even middle households in these places are at risk of long term disruptions to their livelihoods.

(d) Households with inferior housing are particularly vulnerable to hazards: the house itself may be inflicted with costly damage and household assets, equipment and belongings are poorly protected. Housing presents a great challenge to the destitute and poor-to-middle population segments above. Frequently their underprivileged position also strikes through in unfavorable living spaces, such as lower river bank locations, where the flooding of the entire house seems almost inevitable. Thereby, this category is really a function of the conditions referred to in the above categories (a) and (b).

7 Conclusions

As expressive of the performance of the centralized Party-state, rural poverty levels carry huge symbolic value, and therefore tend to be politically construed. It is debatable whether they are a reasonable reflection of poverty: both perceived poverty and calculation based on international standards would produce considerably higher levels. The present research has shown that large segments of the rural population live in social and economic conditions that make them vulnerable to natural hazards. Public support is scant and subjected to the local administrations' varying practices and efficiency. Natural hazards like extreme weather events, possibly related to climate change, have not upset the general economic growth and rising average living standards in the region. Rather, they have accentuated existing inequalities, both among individual households and between coastal and inland localities.

When typhoons and floods strike all communities and all income groups may be affected, though in very diverse ways. The poorer households, some of which live in highly exposed locations, risk losses beyond their coping capacity. With higher incomes, the value of losses also tends to rise steeply: the wealthier households may not be materially threatened, but suffer financial losses that not only impact themselves, but potentially hit back on local areas in terms of loss of employment opportunities. Risk-prone economic activities are evidently on the rise along with Vietnam's embrace of capitalism and high economic growth, including acacia plantation forestry and aquaculture. They are driven by the huge profits that may be had in successive years without disaster-related losses. This shows that despite the risk-prone natural conditions in Central Vietnam, it offers great economic opportunities at the same time. Being left to private initiative and the market, many practices may be described as maladaptive.

Another grave challenge to local-level adaptation relates to the Party-state's tight grip on local organizing, which remains a crucial instrument for policy implementation and propaganda. First of all, a greater space for civil society development, including spontaneous and intentional organizing in communities, might bolster the efficient and accountable use of external inputs, such as for poverty alleviation, disaster relief, local adaptation, and environmental protection. At present, social and 'community' assets are poorly utilized. A gradual transfer of responsibility and funding to community or civil society based organizations might encourage small-scale projects for local development as well as activate some of the surplus labour that evidently exists in the villages. Second, while the present power structure in Vietnamese society inhibits the development of 'community', it simultaneously allows the existence of informal spaces in the extension of formal hierarchies, where illicit practices thrive. These practices have developed in rural education, health care and land management to the point where they seriously deepen the social differentiation already recognized in official statistics, as much as they threaten hard-won gains in poverty alleviation. For genuine 'pro-poor' development, civil society organizing against corruption, inequality, land-grabs, state-centered development and more would be essential elements.

7.1 Methodology in Brief

The survey, including a total of 166 households, covered two coastal district (Hoi An and Nui Thanh), two intermediate level districts (Que Son and Dai Loc) and a mountainous district (Bac Tra My). In each district several communes were chosen as representative of various landscape and land use types. The questionnaire included a range of questions on household income and livelihoods, land use practices, endowments and entitlements, recent events and changes, local organizations and institutions, historical memories about climate and environmental interrelationships, local community responses to earlier climate events, forms of adaptation to other external risks and shocks, and other aspects. Questions mixed quantitative and qualitative forms. The interviewed households were randomly selected across income groups and household representatives were equally divided between men and women. The survey was accompanied by local participatory appraisals, follow-up interviewing to clarify answers or to develop aspects for research, and by various kinds of observation over a lengthy period of time.

References

Adger WN (1998) Indicators of social and economic vulnerability to climate change in Vietnam. CSERGE Working Paper GEC 98-02

Adger WN, Lorenzoni I, O'Brien KL (eds) (2009) Adapting to climate change: thresholds, values, governance. Cambridge University Press, Cambridge

Bankoff G (2003) Constructing vulnerability: the historical, natural and social generation of flooding in metropolitan manila. Disasters 27(3):224–238

Brown M (2012) UN survey questions Vietnamese about corruption. Voice of America (voanews.com) May 3

Bruun O (2012) Sending the right bill to the right people: climate change, environmental degradation and social vulnerabilities in Central Vietnam. Weather Clim Soc 4(4):250–262

Chambers R (1995) Poverty and livelihoods: whose reality counts'. IDS discussion Paper 347

Chaudhuri S et al. (2002) Assessing household vulnerability to poverty from cross-sectional data: a methodology and estimates from Indonesia. Columbia University, Department of Economics, Discussion Paper 0102-52

Coxhead I, Vu HL (2012) The effects of global shocks on poverty and food security in Vietnam (revised draft). Agricultural and Applied Economics. University of Wisconsin-Madison, Madison

Eriksen S et al (2011) When not every response to climate change is a good one: identifying principles for sustainable adaptation. Review article. Clim Dev 3:7–20
EMW (East Meets West Foundation) (2009) Climate Change Adaption Survey. http://downloads.eastmeetswest.org/pubs/emw_climatechangeadaptsurvey.pdf
Gaillard JC (2007) "Natural" disaster? A retrospect into the causes of the late 2004 typhoon disaster in Eastern Luzon, Philippines. Environ Hazards 7(4):257–270
Herrador JJA (2012) The political economy of pro-poor policies in Vietnam: note on the role of civil society in the reduction of school fees for primary education. World Bank Institute (info.worldbank.org/etools/docs/library/.../Civil SocietyVietnam3.pdf
Kang W, Imai KS (2010) Pro-poor growth, poverty and inequality in rural Vietnam. Economics discussion paper series, EDP-1011, University of Manchester
Klump R (2004) Was growth pro-poor in Vietnam?: assessment and policy recommendations. University of Frankfurt and GTZ
LAV (2009) Numerous hydroelectric plants built in central and central highlands regions have made floods in the storm-hit area more serious LookatVietnam.org, 12 Dec
Marino E, Ribot J (2012) Special issue introduction: adding insult to injury: climate change and the inequities of climate intervention. Glob Environ Change 22:323–328
McElwee P (2010) The social dimensions of adaptation to climate change in—Vietnam. The World Bank, Discussion Paper no. 17, Dec
Moser C (1998) The asset vulnerability framework: reassessing urban poverty reduction strategies. World Dev 26(1):1–19
O'Brien K, Leichenko RM (2000) Double exposure: assessing the impacts of climate change within the context of economic globalization. Glob Environ Change 10:221–232
Oxfam (2008) Climate change, adaptation and poor people. Oxfam, UK
Rambo T (2005) Searching for Vietnam: selected writings on Vietnamese culture and society. Kyoto University Press, Kyoto
Schipper L (2007) Climate change adaptation and development: exploring the linkages. Tyndall centre for climate change research, Working Paper 107
Singh M et al (2003) Making democracy work for pro-poor development. Commonwealth Secretariat, London
Sperling F et al. (ed) (2002) Poverty and climate change: reducing the vulnerability of the poor through adaptation. African Development Bank
Takashi O (2009) Rural credit and community relationships in a Northern Vietnamese village. Southeast Asian Stud 47(1):3–30
Thuc VH (2007) Land reform and rural livelihoods: an examination from the uplands of Vietnam. Humboldt-Universität zu Berlin, Dissertation
UNDP (2011) Participatory civil society assessments: experiences from the field. UNDP, New York
VASS (Vietnam Academy of Social Science) (2011) Poverty Reduction in Vietnam: Achievements and Challenges, Hanoi
WB (World Bank) (2010) Poverty headcount ratio at $1.25 a day (PPP) (% of population). http://data.worldbank.org/indicator/SI.POV.DDAY
WB (World Bank with Danida and Sida) (2011) Recognizing and reducing corruption risks in land management in Vietnam. National Political Publishing House, Hanoi
Wisner B et al (2004) At risk: natural hazards, people's vulnerability and disasters, 2nd edn. Routledge, London

Natural Resource Management Impact on Vulnerability in Relation to Climate Change: A Case in a Micro-Scale Vietnamese Context

Tobias von Platen-Hallermund and Anton Mikkel Thorsen

Abstract This chapter analyses the relationship between natural resource management practices and vulnerability in the process of adapting to climate change in Tam Thanh, a rural commune in mid-Central Vietnam. The concept of vulnerability is examined and divided into physical and social vulnerability. Physical vulnerability responds to external stress from natural hazards exacerbated by climate change. Social vulnerability is the ability of a given group or individual to cope with and adapt to any external stress. Based on a household survey, the dynamics of social vulnerability are analysed through the Environmental Entitlements Framework and the Sustainable Livelihood Approach. We find that due to spatial and socio-economic conditions, there are great variations in access to natural resources and endowments which exist between households in the commune and which result in unequal levels of social vulnerability. These differences are reinforced by the impacts of climate change, which are characterised by increased intensity and frequency of typhoons and heavy rainfalls as well as prolonged periods of water scarcity. Natural resource management practices also influence the level of social vulnerability. Poorer households dependent on rice production are especially vulnerable to water scarcity, whereas wealthier households endowed with large plantations of production forest are affected by typhoons. A diversification of income sources reduces the level of social vulnerability. However, alternative income sources beyond climate sensitive natural resources are limited in the commune. The study concludes that a reduction of social vulnerability is essential for a successful adaptation to climate change.

Keywords Vulnerability • Natural resource management • Access • Adaptation • Resource dependency • Climate change

T. von Platen-Hallermund
Sustainable Development Department, World Bank, Washington, DC, USA
e-mail: tvonplaten@worldbank.org

A. M. Thorsen (✉)
Park Service, Copenhagen commune, Copenhagen, Denmark
e-mail: antonthorsen@hotmail.com

O. Bruun and T. Casse (eds.), *On the Frontiers of Climate and Environmental Change*, Environmental Science and Engineering, DOI: 10.1007/978-3-642-35804-3_9,

1 Introduction

Impacts of climate change pose significant challenges, especially for the developing world (Leary et al. 2008; Schipper and Burton 2009). Here, the presence of high levels of vulnerability have been linked to a range of factors, including high dependence on natural resources, lack of equity in terms of access to these resources, weak institutional and financial capacity as well as high rates of poverty (IPCC 2001; Leary et al. 2008; Tomas and Twyman 2005). Vulnerability may relate to a physical and environmental threat caused by changes in weather patterns and climate. But vulnerability is also a social condition, shaped by prevailing economic and institutional contexts, natural resource management and the distribution of resources. Hence vulnerability is unevenly distributed among a population (Chaudhry and Ruysschaert 2007; Adger 1999). Adaptive measures tend to concentrate on reducing the physical determined vulnerability without paying much attention to the issues that control the social vulnerability. In this way, the process of adaptation solely involves technical solutions that maintain existing social vulnerability structures and does little to promote the fundamental social changes needed to effectively adapt to climate shocks and environmental threats (Schipper 2007; O'Brien et al. 2004). Reducing the socially embedded vulnerability implies to address the fundamental, underlying factors that make coping with the impacts of a changing climate difficult. Consequently, it is considered a prerequisite for an adaptation process to take place. This vulnerability approach advocates that the challenge of climate change needs to be addressed in a more holistic way and suggests, that the political, socio-economic and environmental issues, that causes vulnerability should also be addressed, even if they seem to have little or nothing directly to do with the climate (Schipper 2007).

Developing countries' high level of vulnerability is often a result of the dependence on climate sensitive ecosystems for a high proportion of their economic activities and livelihoods (Thomas and Twyman 2005; Abramovitz et al. 2001). Climate change is causing stress and pressure on these ecosystems, forcing natural resource-dependent communities to adapt to new environmental conditions. Here, poverty tends to increase vulnerability determined by geographic and socio-economic factors. For example, poor groups of people in a society are often forced to live in more disaster prone areas on marginalized lands, making them even more exposed and vulnerable to risks such as flooding and typhoons. These marginal areas also have higher marginal costs of access and are often located far from assistance and resource support from government. In addition, intensive and often unsustainable land-uses severely exacerbate the vulnerability to climate change. Very often, natural resources in these areas are in a degraded state, which in turn increases the vulnerability to climate change of both the resources themselves and the people, who are dependent on them (Tompkins and Adger 2004; Adger 1999; Leary et al. 2008). This concern and the arguments behind it, call for a closer examination of how natural resource management influence vulnerability in relation to climate change.

As a consequence of marginalisation and the absence of economic security and access to resources, poor people in Vietnam are highly vulnerable to the impacts of climate variability and climate change (World Wide Fund for Nature 2005a). Therefore, a key issue is the identification of the underlying causes of vulnerability of these poor communities. As more than 90 % (2002 figures) of poor households are located in rural areas in the country (Care 2004, p. 1), an approach that embraces both rural livelihoods and the causes of vulnerability seems relevant. Furthermore, given rural poor households' dependence on agriculture, forestry and other uses of natural resources, it is necessary to focus on the role of natural resource management in the search for ways to reduce the level of social vulnerability (Care 2004; Oxfam 2008).

The inland commune of Tam Thanh located in Quang Nam Province illustrates perfectly the great challenges people living in rural areas face as a result of climate variability and climate change. The area's main source of income is forestry, whereas outputs from agriculture play an essential part in obtaining food security (Household Survey). In this region, climate change—in the form of more frequent and extreme typhoons and changing rainfall patterns—have already affected the livelihoods of the people of Tam Thanh (Phan et al. 2010; Institute of Strategy and Policy on Natural Resources and Environment 2009). However, existing socio-economic inequalities and geographical factors seem to lead to dynamic vulnerabilities within the commune border, which results in differentiated impacts of climate change among the population. By investigating the relation between vulnerability and natural resource management in the small commune of Tam Thanh, we seek to reveal the causes of the uneven vulnerabilities at a micro level and thereby facilitate a successful process of adapting to climate change for the rural poor (Adger 1999; World Wide Fund for Nature 2005b).

2 Vulnerability: A Conceptual Discussion

The literature on vulnerability in relation to environmental stress and natural hazards is extensive and growing, since the concept plays an important role in assessing the adverse impacts of climate change and the possible adaptive measures, that can be taken to reduce the magnitude of this threat (e.g., Adger 1999; Kelly and Adger 2000; Abramovitz et al. 2001; Brooks 2003; O'Brien et al. 2004; Adger 2006; IPCC 2007; Leary et al. 2008). This chapter defines social vulnerability as a state determined by social factors that exist in a community before it encounters a hazard (Cutter et al. 2003; Brooks 2003). Rather than being defined by the likelihood of future adverse impacts from natural hazards, social vulnerability is the inability of a community to cope with external pressures and shocks, and in this case also climate change. Consequently, social vulnerability encompasses all the aspects of a community independent from possible hazards. These factors influence the outcome of the hazard and determine which groups will suffer most as a consequence of this event (Brooks 2003). This interpretation of vulnerability understands it as a socially constructed state (Abramovitz et al. 2001; Adger and Kelly 1999).

Social vulnerability may be defined as: "*...the state of individuals, groups, or communities defined in terms of their ability to cope with and adapt to any external stress placed on their livelihoods and well-being*" (Adger and Kelly 1999, p. 254). This definition encapsulates the need to focus on the stress produced by external changes and the fact that vulnerability is not only caused by climate change, but can be caused by all sorts of changes. Moreover, focus is shifted away from the physical aspects of external stress factors and towards the social and political aspects of vulnerability. Hence, social vulnerability represents a starting point in a vulnerability assessment that aims to understand the causes and distribution of vulnerability before the event of a natural hazard. Furthermore, the purpose is to identify the most vulnerable members of a community and the geographical and socio-economic variations in vulnerability.

The political and socio-economic processes of marginalisation and inequality play a significant role in assessing social vulnerability. Factors such as poverty, inequality, resource dependency, access to these resources and diversification of income sources are considered the most significant indicators of social vulnerability (Adger 1999; O'Brien et al. 2004; Brooks 2003). It is, however, important to note that this understanding of vulnerability recognises the physical conditions as highly influential on the level of social vulnerability, and that environmental and social changes are interlinked. An assessment of the social vulnerability of a community must therefore also examine the characteristics of the physical vulnerability of this geographical area (O'Brien et al. 2004; Adger 2006).

Another key aspect of social vulnerability is the dynamic dimension of the term. It is a dynamic, constantly changing entity, because the social processes and the physical conditions that shape social vulnerability are themselves dynamic. Thus the level of social vulnerability will vary with the seasons (e.g., rainy season versus dry season) or between years, mainly due to the impact from the weather and other external factors such as economic and political circumstances. Consequently, this dynamic aspect of social vulnerability is the most important to capture in an assessment, rather than any measure of vulnerability at a particular point in time (O'Brien et al. 2004; Adger 2006).

The environmental entitlements approach (Mearns et al. 1997; Leach et al. 1999) is a helpful tool to analyse the variation in social vulnerability and the possible options for different natural resource management practices of social groups. Based directly on the indicators already mentioned above, this approach considers the extent to which individuals, households or communities are 'entitled' to make use of different resources. Entitlements are defined as: "*...the actual and potential resources available to individuals based on their own production, assets or reciprocal arrangements*" (Adger 2006, p. 270). Within this understanding social vulnerability is seen as lack of entitlements and an examination of the reasons for this vulnerability will consequently focus on the availability, access and distribution of entitlements as well as how they are changed over time. The political and socio-economic factors which determine the level of social vulnerability also define how the pattern of access to resources is constructed, and this construction is termed the 'architecture of entitlements'. This construction is essential to understand in order to assess the dynamic level of social vulnerability of a household or community. Within this framework, vulnerability to natural hazards and climate change occurs when people, due to political and socio-economic factors, have insufficient entitlements, and when there is a significant loss in the amount or quality of these entitlements (Adger and Kelly 1999; Kelly and Adger 2000; Adger et al. 2003; Adger 2006).

The Environmental Entitlement Framework (EEF) seeks to explain and understand how social and ecological dynamics influence the natural resource management of a diverse group of people in a community and how this management produces particular types of environment. In contrast to the Community-Based Natural Resource Management approach, in which a community is seen as relatively homogeneous with common characteristics different from those of "outsiders", the EEF regards a community as an unit often with divergent areas of interest and internal power relations. Thus, the concept of community as a static unit with mutually dependent people sharing common interests must be modified and altered to a more nuanced perception. Communities are "*composed of people who actively monitor, interpret and shape the world around them*" (Leach et al. 1999, p. 229), and one must keep this in mind when analysing the management of natural resources in a given community. Gender, wealth, age, origin and other socially, historically and culturally determined factors all contribute to social dynamics and may act as barriers in obtaining consensus within a community. Nevertheless, as there may be several examples of relatively socially coherent communities sharing common interests and promoting equity and consensus when trying to achieve shared goals, the notion of "community" still persist. Whether or not a community appears coherent and unified also depends on the scale of analysis. A divergent and divided group of people may appear as an united community to the national authorities, but at a much lower scale of analysis differentiated areas of interest become visible. As the scale of analysis of this paper is on the commune and household level, an understanding of the possible diverging interests and power relations of a given community is crucial, and these social differences and their implications must be taken into consideration (Mearns et al. 1997; Leach et al. 1999) (Fig. 1).

3 The Study Area

The commune of Tam Thanh is located in the central part of Vietnam between 15°23′51″–15°29′31″ north and 108°28′59″–108°34′46″ east in the midland region of Nui Thanh district in the southernmost part of Quang Nam province. Tam Thanh is one out of 16 communes in Nui Thanh; a district which has a population of 147,065 and covers an area of 533.96 km² (Quang Nam Statistical Yearbook 2009). The commune stretches approximately 5 km from east to west

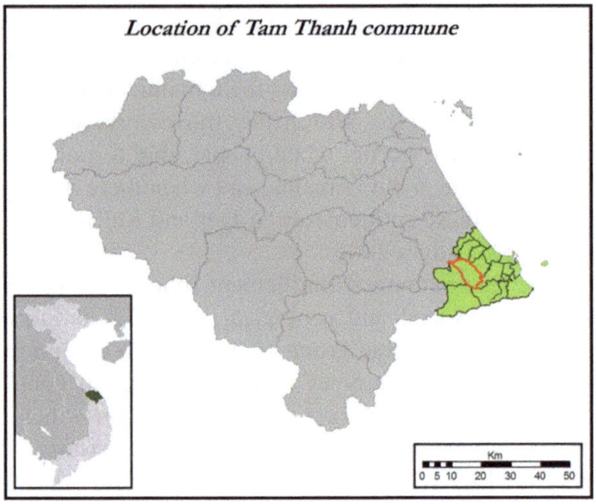

Fig. 1 Location of Tam Thanh commune. Quang Nam province indicating Nui Thanh district in *green* and Tam Thanh commune outlined in *red*

and 10 km from north to south with its eastern border being around 4 km from Highway 1,[1] 10 km from the sea and approximately three hours by car from Danang, which holds one of the main ports of Vietnam. The total area of the commune is 53.93 km² with an elevation from 30 to 300 MASL, and there are 1,058 households totalling 4,168 people divided into four different thons[2] (villages). Two large reservoirs (Phu Ninh and Thai Xuan) are located in the northern and southern part of the commune and provide water to Tam Kỳ, the provincial capital, and to Chu Lai Industrial Zone[3] respectively. In Vietnam forest is divided into three different categories (protection forest, production forest and special-use forest) and the commune holds a large area of production forest and protection forest (within the Phu Ninh Protected Area) while there is no special-use forest in the commune.

Based on interviews and a household survey in Tam Thanh, it becomes obvious that the most important income sources derive from acacia production and casual work, mainly the planting, managing and harvesting of acacia or work in Chu Lai Industrial Zone. Rice production is for own consumption, and the cassava production is mainly used as forage for buffaloes, cows and pigs. The commune has a relatively high percentage of poor with nearly 26 % in comparison to the general poverty rate of 16 % for the whole country (gso.gov.vn). There are, however,

[1] Highway 1 is main national highway with a total length of 2,300 km running north to south.

[2] In Vietnam the commune is the lowest administrative unit. Below this unit, thons exits to make general management and public information easier. Thons are not formal and have no official boundaries but people know about them and they usually have an elected representative that is the connection between the population and the commune authorities.

[3] Chu Lai is located along Highway 1 in Nui Thanh district and consists of an open economic zone (270 km²) and an industrial zone (8 km²).

significant income differences between the households (HH) of the thons in the commune as seen in Table 1.[4]

3.1 Thon 1: Phuoc Thanh

Located in the southern part of the commune, this village is relatively rich as it has the second lowest percentage of poor households and second highest percentage of middle households (see Table 1). The most important income sources are forestry and casual work, and apart from acacia and rice production several households cultivate other types of trees as part of government-support programmes (Household Survey).[5] The thon has the privilege of the main road running through the area, as well as a paved road which runs a few kilometres into the valley, contributing to an enhanced level of infrastructure.

3.2 Thon 2: Trung Hoa

This thon is the smallest, but also most populated thon in the commune. The office of the commune authorities is located here as are a small market and a few local restaurants situated just next to the main road that runs straight through the thon. Apart from this paved road, the thon has a fairly comprehensive dirt road network and with almost no houses located further than around 2 km from the main road, access to the small market, Highway 1 and Chu Lai industrial zone is relatively good. Income is mainly derived from acacia production and

Table 1 Population figures, Tam Thanh 2009

	Total population	%	Total HH	%	Poor HH	%	Nearly poor HH	%	Middle HH	%
Thon 1	1,086	26	270	26	57	21	70	24	143	29
Thon 2	1,176	28	297	28	83	30	80	27	134	27
Thon 3	862	21	206	19	97	36	85	29	24	5
Thon 4	1,044	25	285	27	36	13	60	20	189	39
Total	**4,168**	**100**	**1,058**	**100**	**273**	**26**	**295**	**28**	**490**	**46**

Source Commune authorities

[4] In Vietnam the state categorise all people as poor, middle or rich. However, in Tam Thanh, the rich category does not exist and instead there is a category named nearly poor. As a consequence middle income is the highest official category in Tam Thanh commune.

[5] Household Survey refers to the 36 36 household surveys conducted by the authors (see Sect. 3.4 for further details).

business activities such as transportation of trees and selling of food and groceries (Household Survey).

3.3 Thon 3: Truong Thanh

Thon 3 is located in the northern part of the commune and is by far the poorest thon with only 12 % of the households categorized as middle (see Table 1). There is a large area of 1,338 ha of protection forest (Interview with Commune Officials) as well as the Phu Ninh reservoir, which takes up a large area of the thon and has significant influence on the natural resource management in the area. People with rice fields adjacent to Phu Ninh only have one growing season because high water levels in the reservoir during the rainy season leads to the flooding of their fields. Moreover, the thon suffers from not having a paved road and the 6 km of dirt road that connects the area with the main road is often inaccessible during the rainy season. All this lowers the level of infrastructure. The main source of income is casual work, but also acacia production, and to a lesser extent breeding of buffaloes, cows and pigs, contributes to the economy of the households (Household Survey) (Fig. 2).

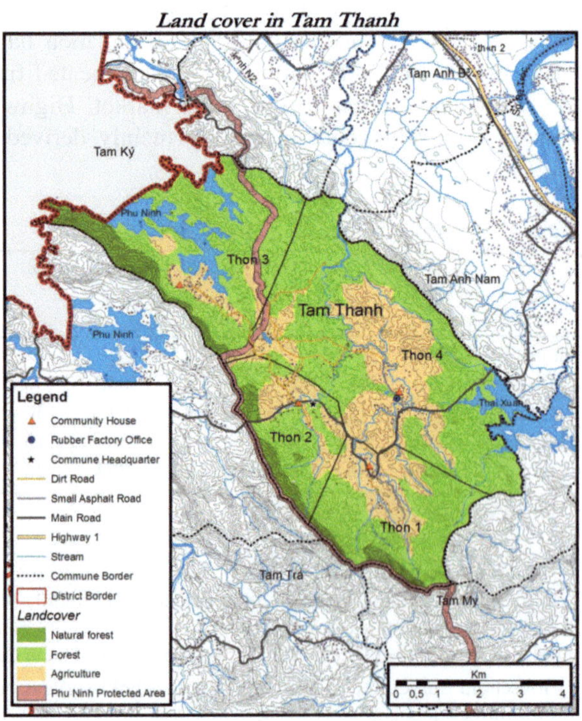

Fig. 2 Land cover in Tam Thanh (2005). The forest category includes both production forest (mainly rubber and acacia) and planted forest. The Phu Ninh protected area is the land west of the indicated line and covers several communes

3.4 Thon 4: Duc Phu

The biggest and richest thon in the commune is thon 4. Located in the eastern part of the commune close to Highway 1, this thon enjoys a relatively high level of infrastructure exemplified by the paved road that runs northward connecting the major part of the population with the main road. The Quang Nam Rubber Group has its office here, as well as the 980 ha rubber plantation leading to a considerable decrease in the area of available agricultural land left for the people in the thon to cultivate. The commune's other reservoir, Thai Xuan, is also located here but does not seem to affect the natural resource management in the thon, mainly because no rice fields are located close to the reservoir. Incomes are primarily derived from forestry and salary from managing the rubber (Household Survey).

The collected empirical data is based on a household survey conducted in 36 households[6] and carried out in the commune in the period from February to April 2010. To investigate the local differences in vulnerability, the fieldwork focused on the four thons and the differences between them, and the 36 household interviews were equally distributed among the four thons. The survey contained both quantitative and qualitative questions, and brought specific data about personal information, socio-economic classification, impacts from natural hazards and climate change, differences in the livelihood capitals, and their natural resource management practices. The sampling strategy was both spatial and non-spatial and as a consequence, geographical location and income category were the main criteria for the selection of households to be interviewed. In both the spatial and non-spatial sampling a stratified sampling approach was used. Furthermore, focus group discussions in each of the thons were carried out, to identify main threats and obstacles for natural resource management in the area. This was done to relate climate change to other obstacles to successful natural resource management. The data was triangulated with semi-structured interviews with key informants from both international NGOs operating in Vietnam and from local authorities at district and commune level. This strategy aimed at producing qualitative information about the underlying causes to vulnerability on both the commune and household level.

4 Physical Vulnerability and Climatic Conditions

The province of Quang Nam is characterised by a tropical and monsoon climate and suffers from water shortages and saltwater intrusion during the dry season (February–July) and flooding and typhoons in the rainy season (August–January)

[6] The more comprehensive survey made for the entire Quang Nam Province finalized 166 household interviews.

(Son et al. 2009; Chaudhry and Ruysschaert 2007). The adverse impacts from extreme weather events and climate variability are highly determined by the geographical characteristics, even in a small area such as that of Tam Thanh commune. The location of plots, land-use, level of slope, and the soil condition are all important factors in determining the physical vulnerability of the different households. Because of this, each specific part of Tam Thanh commune is to a large extent affected by certain natural hazards and not by others. Nonetheless, three main natural hazards generally have an impact in all parts of the commune: floods (heavy rainfall), water scarcity and typhoons.

The vast majority of the annual rain (83 %) falls from August to January and floods occur during this period. Especially October and November receive large amounts of rainfall, mainly caused by typhoons from the South China Sea, and this leads to flooding, especially in the small valleys where water from the mountains is accumulated (Son et al. 2009; Institute of Strategy and Policy on Natural Resources and Environment 2009). Rice production in these valleys is very vulnerable as heavy rain in August/September may coincide with the final period of the summer season, forcing people to harvest the rice before its maturity date. Also, rice plots close to the Phu Ninh reservoir flood during the rainy season when the water level in the reservoir increases as a result of the closing of the lock that controls the water flow in the Tam Kỳ River. As a consequence, the households with rice plots adjacent to Phu Ninh in thon 3 only have the winter-spring growing season, resulting in reduced environmental entitlements. In this way flooding is directly responsible for land scarcity, because farmers need to find more suitable land to be able to generate a sufficient amount of rice. Floods also have severe impacts on the physical capital in vast areas of the commune. Due to the very modest network of tarmac roads, the population that lives in areas with only dirt roads experience decreased mobility. During the rainy season, these dirt roads are very difficult to use and transportation time increases significantly. In spite of these obvious adverse impacts from heavy rainfall, only 5 % of the households (Household Survey) consider floods an important external factor influencing their livelihoods. Instead, the typhoons that coincide with heavy rainfall are regarded as having a dramatic impact on the people of Tam Thanh.

Typhoons regularly hit the commune between August and November. The period of typhoon landfalls has been more difficult to predict as it has gradually shifted from August/September to October/November from the 1950s to the 1990s (Chaudhry and Ruysschaert 2007; Monre 2007; Institute of Strategy and Policy on Natural Resources and Environment 2009), and this is a contributing factor in the change of the start of the winter-spring rice season from late November to late December. Moreover, typhoons have increased both in intensity and impact (Phan et al. 2010; Son et al. 2009; Monre 2007) and are by far the most disastrous type of natural hazards affecting the commune. An important aspect of typhoons is the fact that it is almost impossible to reduce the physical vulnerability to this weather phenomenon, unlike the effects of water scarcity and floods. This is emphasised by the fact that 89 % of the households mentioned typhoons as the single most important external factor affecting their livelihoods over the last 10 years (Household Survey).[7]

During the dry season many areas in Tam Thanh commune suffer from lack of water. Water scarcity is an occasional problem during the rainy season as well, because rainfall patterns have changed during some periods. In general, the water reaches its highest level in November–December and the lowest in May–June and in this latter period, large parts of the commune lack water despite the presence of the two big reservoirs, Thai Xuan and Phu Ninh, as they are primarily used to provide water for Tam Kỳ City and Chu Lai industrial zone (Son et al. 2009; Quang Nam Statistical Yearbook 2003, 2004 and 2009). Because of low physical capital (in this case pumps) and the absence of legal access rights to water, several rice plots in thon 3 lack water despite the fact that the fields are located just next to Phu Ninh reservoir. As a result of, this thon 3 has a double exposure to climate variability with severe impacts on the rice production from both water scarcity and floods leading to a very low yield in several cases (Household Survey).

However, the adverse impact from water insufficiency on the production of rice is also a problem for the other parts of the commune. 33 % of the interviewed households consider water scarcity an important external factor influencing their livelihood strategies and especially the rice production. The main reasons for water scarcity are an insufficient and unequal distribution of the small reservoirs in the commune, which are all located in thon 1 and 2. In addition to this, a change in land use from natural forest to monoculture has resulted in faster runoff and thereby aggravates seasonal shortage of water. Many respondents see lack of water as the main obstacle to achieving a sufficient yield of rice, and even under normal weather conditions the fields suffer from water scarcity in both rice seasons. Moreover, according to the respondents, the periods of low levels of available water are becoming more intensive and last longer. This could be a consequence of the growing variation in rainfall and may also be due to increasing evaporation triggered by rising temperatures (Phan et al. 2010; Chaudhry and Ruysschaert 2007). This absence of rain also results in problems concerning the planting of acacia. Acacia is generally very resistant to water scarcity, except during the planting period where it is highly dependent on water. For the people with acacia plantations, who were highly affected by the typhoon Ketsana in September 2009 and consequently in great need of planting new trees, the absence of rain was yet another climatic factor affecting the acacia plantations at different times during the year.

5 Natural Resource Management Practices

During the Vietnam War (1955–1975) the district of Nui Thanh was strongly affected by combat and violence and was therefore partly abandoned, until people began to return to the area from 1975 and onwards. In that period, big areas of

[7] The respondents were asked to name the most important external factors impacting their livelihoods within the last 10 years. Typhoons were ranked first with 89%, water scarcity came second with 33% and third with 19% was disease in family.

natural forest in Tam Thanh were cleared and subsistence agriculture started along with tea production managed by Duc Phu Company (a state-owned company), which covered an area of 980 ha. During the 1980s, this area was partly transformed into pineapple production, but this was abruptly halted in the beginning of the 1990s because of the collapse of the biggest purchaser of pineapple, The Soviet Union. From then on, the state-owned Quang Nam Rubber Group (QNRG), which then took over the land of Duc Phu Company, started planting rubber trees and from 2009 to 2010 this also included acacia. QNRG, an affiliate of the state-owned enterprise Vietnam Rubber Group, is by far the largest landowner in the commune with its 2,000 ha all located in thon 4. Today rubber is widely grown in Tam Thanh, but it all belongs to QNRG and the company is therefore the only producer of rubber in commune. Yet it has a great influence on the households, as many work for the company, managing the rubber trees and collecting the rubber. In total, 350 people have obtained long-term contracts at QNRG and the company is in fact planning to expand in the coming years. Hence more people can be expected to be employed by the QNRG in the future. Rubber can be collected 10 months a year, while for two months the trees have to regenerate, and it is therefore a stable income for the contract-holders. Furthermore, it lessens the impacts from stresses and shocks that may be experienced with respect to their alternative income sources, thus reducing social vulnerability. Few have expressed desire to start producing rubber themselves, as it is expensive to start up and because it takes eight years before the collection of rubber can begin. So even though the profit is more lucrative, most are reluctant to initiate personal rubber production. Rubber production, like acacia, was hit hard by Ketsana and many trees broke, underlining the risks of such a capital intensive and long-term investment. The aftermath of Ketsana precipitated the introduction of a new technique for the production of rubber, where the trees are cut down to four meters after 3 years, to avoid collapse during typhoons.

The people in the commune also grew pineapples during the 1980s, and cultivated tea and cassava, but they shifted almost all cultivation, apart from paddy rice fields, to acacia in the period of 2003–2005. Today, rice and acacia, which take up the wetlands and primarily the lower parts of the upland respectively, are the two main natural resource management practices of the population. Moreover, cassava is intercropped on the acacia fields on a considerable scale, whereas other crops such as pineapple and beans only are cultivated to a small degree.

Acacia production has generally improved the economic situation in Tam Thanh. Income from sales and the casual work generated from planting, managing and especially harvesting the trees have contributed significantly to the economy of the households and have allowed more time to a diversification of income sources. The level of social vulnerability has in this way been reduced. Nonetheless, the profits received per ton of acacia vary considerably due to the cost of harvesting, which is primarily determined by the acreage of the plantation and the distance to the main road (See also "The Push for Plantations: Drivers, Rationales and Social Vulnerability in Quang Nam Province, Vietnam"). Hence, the economic benefits from acacia production are not equally distributed among the population in Tam Thanh. Relatively large acacia plantations located close to the main road benefit more economically from the sale of acacia than smaller plantations not connected by tarmac road (e.g., plantations in thon 3). Due to this and to the general income differences within the commune, natural hazards such as typhoons impact farmers differently. Where middle income households endowed with a great amount of acacia are highly impacted in absolute economic terms, poorer households with a smaller plot for acacia cultivation, often suffer more as their losses are relatively higher (Household survey). Apart from living in what are often weakly constructed houses, poorer households are likely to be more dependent on income from acacia, and have little or no savings to invest in new acacia production and therefore are more socially vulnerable. On the other hand, wealthier farmers with larger plantations are consequently at risk of economic losses that are potentially much higher, but are less socially vulnerable as they usually have a non-agricultural income and money to restart (see also "The Push for Plantations: Drivers, Rationales and Social Vulnerability in Quang Nam Province, Vietnam").

Almost all households in Tam Thanh produce rice, and the households that do not have a rice field are households categorised as middle income. Households keep their rice fields until they have a secure and sufficient income to buy their rice, which underlines the importance of rice as a subsistence crop in the commune. The past years have shown a reduction in the number of seasons for rice cultivation. Earlier, rice was planted three times a year, as opposed to twice a year now. This is a consequence of a suggestion made by the district authorities in 2000 to avoid the rainy season, which results in higher yields per harvest as the crops are not affected by severe rainfalls. Furthermore, the strategy also reduces the investment load.

Cassava has also played a significant role in Tam Thanh, but in recent years cassava has been supplanted by acacia because of the low income cassava provides. In addition, acacia trees reduce surface water flow more efficiently and thereby mitigate against erosion (Adger et al.2005). A new type of cassava has been introduced that makes it possible to intercrop with acacia the first year, whereas the earlier type of cassava grew too big to allow enough sun to penetrate for the young acacia trees and thereby killing them (Interview with Commune Officials). Today cassava is widely intercropped on the acacia fields in the first year and is mainly used for forage for livestock or is sold and transported to Nui Thanh, where it is sent to factories that produce glutamate. This way cassava provides an extra set of entitlements from the acacia fields.

Fifty-five % of the interviewed households have at least pig, cow or buffalo. For most people buffalos are essential to their livelihood, as the livestock is used to plough the rice fields. It is, however, only 17 % who own a buffalo, presumably because the cost of buying and feeding a buffalo are quite high. More farmers have pigs (22 %) and cows (30 %), generally for the purpose of selling them, while the excrement is used as fertiliser on the fields. However, experiences in recent years have shown considerable problems with diseases, resulting in the loss of livestock. In spite of this fact, 12 % of the respondents have taken loans to buy livestock. This livelihood strategy reduces social vulnerability through income diversification.

The general development in Vietnam has led to an increased environmental degradation (World Wide Fund for Nature 2005a). One consequence is the drop in forest cover from 40 % in the 1960s to 26 % in 1997.[8] The loss in forest cover eventually leads to soil degradation and exacerbates flooding due to increased variability in water flow (Adger et al. 2005). This is also the case in Tam Thanh, where the natural forest cover was almost completely cleared during the 1970s and 1980s. Today, the farmers complain about soil degradation, bank erosion and water scarcity that results in reduced agricultural output. Water scarcity is a major problem for many households that produce rice, as only 20 % have access to the four reservoirs in the commune. This may in part be caused by the more severe dry seasons that the commune experiences. However, the increased variability in water flow due to the loss in forest cover is also a factor. The unequal access to the four water reservoirs in the commune, all located in thon 1 and 2, further exacerbates this situation. Though the farmers view bank erosion as a result of changed weather patterns, causing fewer, but more intense, rainfalls, part of the explanation lies with natural resource management that unfortunately exacerbates erosion, as crop cultivation on the slopes usually does not reduce water flow to the same extent that natural forest previously did.

The shift to acacia trees has nevertheless helped reduce the environmental degradation in the commune. Acacia is a legume and therefore has the ability to fix atmospheric nitrogen and create a favourable environment for many other soil organisms (Van Bueren 2004). It also reduces water flow and minimises general erosion and landslides more efficiently than pineapple and tea, which were mainly grown on the slopes before (Interview with Commune Officials). Furthermore, acacia improves the quality of the soil that is otherwise in poor condition. Cultivating acacia trees, nonetheless, increases impact from typhoons, since these relatively tall trees are more vulnerable to strong winds than pineapple plants, for instance. Furthermore, the perennial cycles in which acacia trees are produced increase physical vulnerability to typhoons. In this way, the self-induced vulnerability has increased along with the expansion of acacia production.

6 Access and Endowments

Households in Tam Thanh have on average 2.4 ha of land, ranging from 1.1 ha in thon 2 to 3.9 ha in thon 4 and with individual differences ranging from 0.02 to 14 ha, though the majority have between 0.5 and 2 ha (Household Survey). When households register their land-use rights, they get a land-use certificate popularly known as the Red Book. Apart from serving as proof of legitimate land tenure the certificates also give access to cheap loans in the Social Policy Bank and thereby facilitate access to financial capital, In Tam Thanh all villagers have a Red Book for their rice fields (annual crops), and most are either waiting

[8] There is however, an ongoing debate about actual forest cover in Vietnam (See Thulstrup et al. for an in dept discussion on this).

for or holding a Red Book for forest land such as the acacia plantations (perennial crops). There are great differences though within the various thons. None of the households in thon 4 have been registered, though some have had their land measured, but are waiting to receive the Red Book certificate. This is because QNRG owns the vast majority of the land in this thon. On the contrary, thon 3 has the highest number of registered households despite it being the poorest and most isolated thon in the commune. The remaining two thons, like thon 3, are quite far in the process of obtaining Red Books certificates for the households. These large differences clarify the diverging interests that exist within the commune and underline the importance of assessing the social differences and diverging needs within the commune.

The process of registering land in the Red Book has lead to a more secure endowment situation and easier access to cheap loans through the Social Policy Bank. This development also seems to have contributed to the shift from pineapple to acacia, with the result that the livelihood of the villagers in Tam Thanh has improved. The change to perennial crops, a more long-term and capital intensive investment, conversely increases the risk of investment. So while general social vulnerability might have decreased due to more secure endowments and better access to financial capital, vulnerability to weather extremes seems to have increased ("The Push for Plantations: Drivers, Rationales and Social Vulnerability in Quang Nam Province, Vietnam"). The large production of cash crops has also created a strong dependence on the market, and consequently the market situation is an essential factor influencing the livelihoods of farmers in Tam Thanh. Volatile market prices are often a significant stress factor for rural people producing cash crops, as the dependence on the market makes them vulnerable to market fluctuations over which they have no control (Adger 1999; Thomas and Twyman 2005). This is particularly the case in Tam Thanh where a large percentage of the income depends on one crop, namely acacia.

The amount of land is not the only factor influencing the endowment situation of a household. There are, for various reasons, significant differences in the quality of the land in Tam Thanh. Even though it is not a significant challenge in the commune, people have on occasion had problems with erosion. When the trees have just been cut and the soil is left bare, it is very vulnerable to erosion and landslides and therefore the acacia trees are almost always harvested in the dry season where the risk of erosion is the smallest. Generally, a connection exists between the slope gradient and the risk of erosion (Folving 2007), but despite the fact that most of the acacia fields are on slopes, there are no significant problems with erosion in the acacia plantations. In addition, people that have land near a stream suffer from bank erosion in the rainy season. This is an increasing problem as rainfall becomes more intense, thereby increasing the pressure on the streams periodically. The general soil quality in the commune has caused constraints in the natural resource management. The farmers from thon 4 complain the most about the poor soil quality, arguing that it was degraded by the earlier tea production by Duc Phu Farm. Nevertheless, the quality of the soil is not a major obstacle with the current land-use, as acacia trees actually improve the soil and can grow in bad soil conditions (FAO.org).

One factor that has a substantial impact on the quality of the land is the Phu Ninh reservoir. Phu Ninh supports Tam Kỳ with energy and water flow control. This in turn entails that if the area around Tam Kỳ experiences severe water scarcity, the locks of Phu Ninh will be opened to sustain a steady flow of water to Tam Kỳ, and thereby the water level in Phu Ninh will drop more dramatically than if it was only due to water scarcity. In times of flooding, the locks will be closed to halt the water flow to Tam Kỳ, resulting in a higher water level. The farmers in thon 3 who live close to the shores of Phu Ninh therefore struggle with the impacts from a change in water level that is more severe because of the sustained water flow to Tam Kỳ (Household Survey).[9] As mentioned earlier, the farmers here have only one growing season for rice, because the area is flooded in the rainy season, which has also led to a shift to a faster growing variety of seeds. This situation affects the transformation of environmental entitlements as the agricultural output of rice derived from endowments is significantly reduced in this thon.

The commune has another big reservoir on its eastern border with Tam Hiệp commune. It supports the industrial zone of Chu Lai but has little influence on Tam Thanh. In addition, there are four smaller reservoirs and 40 ponds. The four smaller reservoirs are all located in thon 1 and 2 and can only support 20 % of the production of paddy rice in the commune where as the rest depends on precipitation. This is because the 40 ponds that are spread around the commune have very low capacity and are destroyed every time it floods. Water is redirected to them before the dry season in order to support irrigation during this period. However, this is not sufficient in itself, and therefore 80 % of the paddy rice in Tam Thanh is rain-fed (Interview with Commune Officials). The 20 % that is sustained by the four small reservoirs have major advantages in comparison to the rest of the households in the commune. After the construction of the four reservoirs, all within the last 10 years, the production of rice has increased, and the lives of the farmers with fields supported by the reservoirs have improved. This is also clear for the majority of the farmers in thon 2 that do not experience water scarcity as they receive water all year round. Consequently, access to water is decisive for the production of paddy rice and is one of the reasons for the differences in environmental entitlements and social vulnerability within the commune.

The commune is well-connected to Highway 1 by an asphalt road built from 2003 to 2005 and financed by QNRG. Nevertheless, access varies widely within the commune as the asphalt road stretches just 11 km and thus reaches only a limited area in the commune; the rest is connected by dirt roads. As the asphalt road runs through thon 1, 2 and 4, these thons generally enjoy a stable and secure connection independent of the weather. Thon 3, however, is located far from the asphalt road and is therefore dependent on a dirt road for access to markets, work, health care

[9] Several studies have assessed the impact from dams in the basin and on the surrounding livelihoods, see e.g., ADB (2008)—Capacity Building in the Strategic Environmental Assessment of the Hydropower.

and schools, etc. The poor accessibility to the main road is seen by the people in thon 3 as the most significant obstacle to improve their livelihood situation (Household Survey), and thus contributes to their social vulnerability.[10]

7 Vulnerability and Adaptation

Land is essential for individual households because the predominant source of income is activities related to natural resource management. In this regard there seems to be a clear link between acreage and income, in that people with the largest area of land have the highest income. Despite the fact that the amount of total land in the commune appears to be sufficient for the population, the unequal distribution of land is a major challenge, which means that some people in many cases lack endowments. This situation perpetuates social vulnerability. The differences in access and endowments between the socio-economic groups are exacerbated by geographical location. Thon 3 lacks water, land and sufficient infrastructure and is also the poorest thon in the commune. As a consequence of all these factors, the most socially vulnerable households are located in this thon. Thon 4 is the richest, has a good infrastructure, and many household members earn extra income working on the rubber plantations. Thon 1 and 2 have the smallest acreage per household, but includes the commune's four water reservoirs and thus has more physical capital to support its agricultural production. While the construction of the main paved road has contributed to general economic growth in Tam Thanh, infrastructure remains insufficient in absolute terms, which reinforces existing inequalities. The areas where the poorest households live typically lack adequate infrastructure, which constitutes a major obstacle to accessing these areas and in turn strengthens the geographical differences of social vulnerability.

Institutional practices have a high influence on the enabling and constraining factors of development that affect inequalities in Tam Thanh. In the past, the state had a central role in providing collective security, but this has changed and today the state has a more marked-based policy (Chaudhry and Ruysschaert 2007). Current policies in the agricultural sector focus on industrialising and intensifying agricultural production, on increasing investments in irrigation and extension services and on promoting agro-forestry (World Wide Fund for Nature 2005a). In Quang Nam this is illustrated by the strong influence that the agricultural extension service has and by the shift to acacia production seen in recent years. This shift has reduced social vulnerability by increasing overall income and by releasing household capacity to focus on other

[10] There is an interesting connection between what the people in Tam Thanh see as the major impacts on their livelihoods in the last 10 years, contrary to the main obstacles in the future. Typhoons and water scarcity have been pointed out as the major impacts. Whereas capital and, for thon 3 the quality of the road connecting the thon to the rest of the commune, have been pointed out as the main obstacles.

opportunities for earning income. In addition, the construction of the asphalt road to the commune has strengthened the integration of Tam Thanh commune with the lowland communes in the district. This might eventually lead to a higher diversification as more people go to Nui Thanh and Chu Lai to buy and sell agricultural products, or perhaps to apply for jobs during periods of low demand for agricultural work.

7.1 Diversification and Resource Dependency

There is a potentially high level of individual vulnerability in communes such as Tam Thanh where income generated from natural resources contributes significantly to the economy of the households (Tomas and Twyman 2005; Adger 1999; Adger et al. 2005). Livelihood strategies in such cases are likely to be very climate sensitive and physically vulnerable to natural hazards such as typhoons, water scarcity and floods. However, diversification of income sources may play an important factor in reducing social vulnerability. Access to multiple types of resources and income earning opportunities increase people's resilience to climate events and other external shocks such as price fluctuations or crop diseases that result in the disruption of particular sources of income (Chaudhry and Ruysschaert 2007; Thomas and Twyman 2005).

As previously mentioned, income from acacia production is by far the most important source of income for households in Tam Thanh. But the tree production has also enabled the population to engage in other income activities, because the production of acacia is significantly less time-consuming compared to earlier natural resource management practices. Consequently, casual work and enterprise also contribute considerably to the economy. Yet, both these activities are linked to some extent to the production of acacia in the form of planting, managing, cutting and transporting the trees. Additional income from this kind of activity is a diversification within forestry and climate sensitive natural resources. For that reason, as income sources, they remain vulnerable to natural hazards (Baumann 2002; Adger 1999; Thomas and Twyman 2005). Diversification beyond acacia in particular, and natural resources in general, are however, less common in the commune. Nonetheless several respondents receive income from activities less dependent or non-dependent on the environment, such as breeding, industrial work, service jobs (restaurants and cafés) or support from the government. This type of diversification is particularly prominent in the wealthier households that receive a significant proportion of their income from non-forestry and non-agricultural activities. In this sense, they are less socially vulnerable to natural hazards compared to poor households that tend to diversify within climate sensitive natural resources ("Livelihood Strategies Under the Constraints of Climate Change Vulnerabilities in Quang Nam"). Having said that, several poorer households diversify through investment in livestock, but due to a recently high loss of especially pigs and cows because of various diseases, this investment seems rather insecure. The importance of diversification is illustrated by the fact that 75 % of the households generate income from two or more activities, but primarily from climate dependent sources.[11] Despite a diversification that remains within the realm of natural resources, households nevertheless show a level of responsiveness to external stress that may be important in adapting to climate change (Tomas and Twyman 2005).

There is no direct link between the number of income sources and the total income, although households with only one source of income are almost exclusively categorised as poor. Moreover what is evident is a spatial tendency of more climate dependent income sources in the remote thon 3 compared to the three other thons. In contrast, thon 2 in particular has a high level of diversification beyond activities reliant on the environment, mainly due to the presence of the market, the commune headquarters, public institutions and the various cafés and restaurants. Income from forestry in thon 2 also contributes less to the household economy than in the other thons, possibly because of the obvious scarcity of endowments in the form of agricultural land as the thon is the most populated and smallest in the commune.

The geographical correlation between the distance to the main road or the urban area east of Tam Thanh and the reliance on climate dependent income sources is an important factor determining the variation in the level of social vulnerability. Diversification options for households living far from the main road and urban areas are generally limited to diversifying within the use of natural resources. On the other hand, households with better access to a high level of infrastructure and closer location to industrial jobs have enhanced possibilities for generating non-climate dependent income. In this regard the population in thon 3 and other remote areas with low physical capital in the commune are more socially vulnerable when natural hazards occur than other thons.

Apart from the geographical variation in the level of social vulnerability, there is an evident link between natural resource management practices and the type of natural hazard that affects households the most. It is predominately the wealthier households, often endowed with the largest acacia plantations that regard typhoons as the most serious obstacle to their livelihood, whereas the poorer households fear water scarcity the most. While typhoons impact farmers with large acacia plantations economically, water scarcity affects poor households highly dependent on rice for own consumption, not in an economic sense but in their ability to generate sufficient food. Despite the fact that poorer households receive a relatively higher proportion of their income from acacia compared to wealthier households, food security is ascribed paramount importance.

8 Conclusion

Various enabling and constraining factors at the community and household level strongly influence the varied level of social vulnerability of the population in Tam Thanh. These factors, shaped by prevailing economic and institutional contexts,

[11] The percentage of households who has two or more income sources corresponds to number in the larger survey made in Quang Nam Province.

agriculture and forestry policies and natural resource management practices, are all dynamic and constantly in a state of flux (Chaudhry and Ruysschaert 2007; O'Brien et al. 2004). The same can be said about the environmental and bio-physical conditions that impact the physical vulnerability, and today the population experiences more frequent and intense natural hazards as a result of climate change. This dynamic dimension of vulnerability is critical to this paper's interpretation of the term, in which social vulnerability is a state that exists within a society before impacts from climate change occur. Hence, in order for a process of adapting to climate change to take place, it is necessary to address the above mentioned factors that currently challenge the progress toward reducing the level of social vulnerability (Schipper 2007).

The population of Tam Thanh has a long history of adapting to environmental and political change, as recently illustrated with the major shift to acacia production. This significant land-use change was strongly encouraged by government policies and the construction of wood-processing factories in the district, and it has had great implications for people's income situation and physical vulnerability. Thus, the elements of social vulnerability that are determined by natural resource management practices are to a large degree a function of external factors beyond the control of the individual households in the commune. Moreover, institutions and government policies also influence other crucial elements of social vulnerability such as inequality, access to resources and the level of infrastructure.

In Tam Thanh, an important reason for the relatively high level of social vulnerability among certain groups is the lack of endowments of land needed to increase acacia and rice production. A more equal distribution of land rights within the commune would reduce the level of social vulnerability of the people that at present suffer from rice shortages or inadequate income from acacia production due to land scarcity. Furthermore, impacts of climate variability and climate change, such as more frequent and intense typhoons and changing rainfall patterns, increase the pressure on the land and consequently reinforce the importance of an equal land distribution.

A structural and more technical solution to this insufficiency in the output from acacia and rice productions would be to raise crop productivity, introduce short-duration and drought-resistant seeds and promote an enhanced irrigation system. More drought-resistant rice seeds that also grow faster have already been introduced to the farmers by the government and supported through training and agricultural extension services. However, water scarcity remains an important obstacle for rice production. Access to water is unequally distributed among the people in the commune. Construction of irrigation systems and smaller reservoirs to ensure a safe and stable water supply during the dry season would reduce both the social and physical vulnerability of the poorer households that are most dependent on food from the rice production.

These concrete recommendations are positive, technical solutions. They focus on solving problems directly linked to climate change while at the same time also reduce social vulnerability on a wider basis. The latter because they simultaneously, and independently from changing weather patterns, increase and stabilise income from existing natural resource management. Such structural measures would, as seen, first and foremost address some of the causes of social vulnerability regardless of the presence of climate change. But as they also reduce the level of physical vulnerability, the process of adapting to climate change is also facilitated, which leads to a mitigation of the adverse impacts from more frequent and intense natural hazards.

Another important cause of social vulnerability is the absence of alternative income sources beyond the use of natural resources. Despite the fact that the shift to acacia production has allowed farmers to use more time on other income activities, non-climate dependent income sources are relatively rare in the commune mainly because of lack of such alternatives in the area. Tam Thanh is located relatively close to the urban area in Nui Thanh and to the industrialisation that takes place in this district and in the province of Quang Nam in general. The opportunities for income diversification are therefore unquestionably increasing. Further infrastructure development in remote areas of Tam Thanh would contribute to better access to alternative income sources beyond the use of natural resources, and also improve the general mobility, not least during the rainy season.

Non-structural measures that ensure equal distribution of land rights and diversification of livelihoods reduce the level of social vulnerability and thereby strengthen the ability to respond to external shocks, which ultimately increase the capacity to adapt to climate change. In effect, the response to immediately changing political and environmental conditions is facilitated and constrained by the same dynamic factors that enable climate change adaptation (Adger et al. 2005). Addressing the causes of social vulnerability to these immediate challenges allows adaptation to uncertainty, which has been increasingly identified as a distinguishing characteristic of climate change (O'Brien et al. 2004; Chaudhry and Ruysschaert 2007; Institute of Strategy and Policy on Natural Resources and Environment 2009; Adger 1999).

The most socially vulnerable households are those that have the least endowments and are most geographically isolated. These are also the households that gain the least environmental entitlements for their endowments and furthermore face the most difficulties obtaining alternative income sources beyond natural resource management. The large differences in endowments, access to natural resources and environmental entitlements that exist in the socio-economic and geographic units in Tam Thanh put constraints on the natural resource management for the most socially vulnerable. These differences are reinforced by the impacts of climate change. An approach that addresses some of these underlying causes will facilitate a process of adaptation to climate change that reaches the most vulnerable households in a community.

References

ADB (2008) Technical Assistance Socialist Republic of Viet Nam:Capacity Building in the Strategic EnvironmentalAssessment of the Hydropower Project. International Centre for Environmental ManagementBrisbane, Queensland, Australia

Adger WN (1999) Social vulnerability to climate change and extremes in coastal Vietnam. World Dev 27(2):249–269

Adger WN (2006) Vulnerability. Tyndall Centre for Climate Research, Norwich

Adger WN, Kelly PM (1999) Social vulnerability to climate change and the architecture of entitlements. Mitig Adapt Strat Glob Change 4:253–266

Adger WN et al (2003) Adaptation to climate change in the developing world. Prog Dev Stud 3:179–195

Adger WN et al (2005) Successful adaptation to climate change across scales. Glob Environ Change 15:77–86

Abramovitz JN et al (2001) Adapting to climate change: natural resource management and vulnerability reduction. Background paper to the task force on climate change, adaptation and vulnerable communities, on and vulnerable communities, Retrieved April 24, 2012, from http ://www.iisd.org/pdf/2002/

O'Brien K, et al. (2004) What's in a word? Conflicting interpretations of vulnerability in climate change research. CICERO working paper 4, Oslo

Baumann P (2002) Improving access to natural resources for the rural poor - A critical analysisof central concepts and emerging trends from a sustainable livelihood perspective. Food andAgriculture Organisation of the United Nations

Brooks N (2003) Vulnerability, risk and adaptation: a conceptual framework. Working paper 38. Tyndall Centre for Climate Change Research, Norwich

Bueren MV (2004) Acacia hybrids in Vietnam. Impact assessment series Report No. 27, Australian centre for International Agricultural Research, Canberra CARE (2004): Civil action for socio-economic inclusion in natural resource management, Phase 2 (CASI). Program document, January 2004–December 2009, Care International in Vietnam and Care Denmark

Care (2004) Civil action for socio-economic inclusion in natural resource management, Phase 2 (CASI). program document, January 2004–December 2009, Care International in Vietnam and Care Denmark

Chaudhry P, Ruysschaert G (2007) Climate change and human development in Viet Nam. Human Development Report 2007/2008, UNDP, New York

Cutter SL et al (2003) Social vulnerability to environmental hazards. Soc Sci Quarterly 84:242–261

Folving R (2007) Farming system changes among ethnic minorities in the Vietnamese uplands. PhD Dissertation. University of Copenhagen, Copenhagen

Institute of Strategy and Policy on Natural Resources and Environment (2009) Viet Nam assessment report on climate change (VARCC). ISPONRE, Hanoi

IPCC (2001) Climate change 2001, working group ii: impacts, adaptation and vulnerability. Third assessment report of the Intergovernmental panel on climate change. Cambridge University Press, Cambridge

IPCC (2007) Climate change 2007, working Group II: impacts, adaptation and vulnerability. Fourth assessment report of the Intergovernmental panel on climate change. Cambridge University Press, Cambridge

Kelly PM, Adger WN (2000) Theory and practice in assessing vulnerability to climate change and facilitating adaptation. Clim Change 47:325–352

Leach M et al (1999) Environmental entitlements: dynamics and institutions in community-based natural resource management. World Dev 27(2):225–247

Leary N et al (2008) Climate change and adaptation. Earthscan, London

Mearns R et al (1997) The institutional dynamics of community-based natural resource management: an entitlement approach. Paper funded by UK economic and social research council's global environmental change programme, Phase IV, Institute of Development Studies, University of Sussex

Monre (2007) National target program to respond to climate change. Hanoi

Oxfam (2008) Vietnam: climate change, adaptation and poor people. Oxfam International, Oxford

Phan V T et al. (2010) Study on Development of Detailed Climate Change Scenarios for Mid-Central Vietnam. Report on Sub-project 3, Project P1-08 VIE, Hanoi

Quang Nam Statistical Office (2003) Quang Nam statistical yearbook 2002. Statistical Publishing House, Tam Ky
Quang Nam Statistical Office (2004) Quang Nam statistical yearbook 2003. StatisticalPublishing House, Tam Ky
Quang Nam Statistical Office (2009) Quang Nam statistical yearbook 2008. Statistical Publishing House, Tam Ky
Schipper ELF (2007) Climate change adaptation and development: exploring the linkages. Working paper No. 107. Tyndall Centre for Climate Change Research, Norwich
Schipper ELF, Burton I (eds) (2009) The Earthscan reader on adaptation to climate change. Earthscan, London
Son LD et al (2009) Climate change adaptation survey: Quang Nam province, central region. Vietnam East Meets West Foundation, Oakland
Tomas DSG, Twyman C (2005) Equity and justice in climate change adaptation amongst natural-resource-dependent societies. Glob Environ Change 15:115–124
Tompkins EL, Adger WN (2004) Does adaptive management of natural resources enhance resilience to climate change? Ecol Soc 2:10
World Wide Fund for Nature (2005a) Agricultural development in Quang Nam province, Vietnam: assessing policy impacts. WWF-Macroeconomics Program Office, Washington
World Wide Fund for Nature (2005b) Community forest management and protection in Quang Nam province, Vietnam. WWF MOSAIC Project, management of strategic. Area for integrated conservation, Washington

Climate, Environment, and the Role of Media

Catherine McKinley

Abstract This chapter analyses media coverage and public debate on climate and environmental change in Central Vietnam. As both the dissemination of public information and awareness-raising are important elements in climate change adaptation, both conventional media and the new digital media are in focus. Equally relevant, however, are public debates on climate and environmental issues and the extent to which alternative visions and civil society voices have access to old and new media. Since Vietnam is a one-party state, in which the media in principle are controlled by public authorities and are intended to support state policy, certain media have limited credibility among the public. Nevertheless, investigative journalism is on the rise and the public is increasingly seen to use the media as a route to express discontent with land grabs, unequal development, pollution, illegal mining, environmental problems and so forth. Slowly, a trend towards more open media spaces, particularly among the online media, enables both civil society groups to represent disadvantaged people and concerned scientists to express alternative policy options.

Keywords Media • Public information • Climate debate • Civil society • Alternative policies

1 Introduction

Vietnam is a one-party state which until recently allowed only one type of media–the state-owned to operate. But the traditional ties between this state-owned Old Media and the Communist Party are weakening as a result of both carefully engineered economic reforms and more organic social change that is resulting from economic liberalisation and is largely beyond the party's control. At the same time (and for many of the same reasons), there is rapid growth in the online New Media, which is forcing the Communist Party to facilitate political and social debate that a decade ago would have been unthinkable.

C. McKinley (✉)
Hoi An, Vietnam
e-mail: catherine.mckinley@gmail.com

O. Bruun and T. Casse (eds.), *On the Frontiers of Climate and Environmental Change*, Environmental Science and Engineering, DOI: 10.1007/978-3-642-35804-3_10,

The media, both old and new, are for the most part highly regional, with key centres in Hanoi and Ho Chi Minh City where the vast majority of media talent is based. While smaller regional hubs exist in other parts of the country, the expertise, money and other resources needed to operate an effective media are seldom sufficiently available outside of the two major cities. In the central region, most media is based either in Danang or Hue on the coast or in major inland highland towns such as Buon Ma Thuot and Dalat. Many news managers in these places complain of poor resources, low staff capacities, and limited finances.

State-owned news organisations fall into three major categories: (1) national-level Communist Party (CPV) mouthpiece organisations, (2) national, provincial or municipal organisations that are owned by the state but are relatively free of central control, and (3) all other official news groups run by lower-level party or government bodies that are tightly aligned to them but have very little central management or oversight. The first group includes two newspapers—Nhan Dan ('The People') and Quan Doi Nhan Dan ('The People's Army')—broadcast media—Vietnam Television(VTV) and Radio the Voice of Vietnam (VoV)—and the Vietnam News Agency (VNA). The newspapers are owned by the CPV while the others are described officially as 'government attached agencies' undertaking 'non-business activities' such as publishing party guidelines and policies. Their links to the state make them ideal for disseminating information broadly and quickly: they have good access to official information and broad network and distribution coverage (especially state broadcasters).

News organisations that may move away from party-line news coverage are clustered within the second group, particularly within the print and online media sectors (broadcast, because of the cost of developing the physical infrastructure required to operate, remains more tightly aligned with state paymasters). Some larger newspapers are able to facilitate discussion of topics the first category of news groups cannot, but with the exception of a small minority they often lack the official connections and nationwide reporting and distribution networks that the Party's media enjoy. The editors-in-chief of major newspapers are relatively powerful and may sometimes be able to influence policy makers when it comes to media content, but smaller print publications have much less voice, and those in Central Vietnam—beyond the centres of Hanoi and Ho Chi Minh City—are extremely unlikely to be able to push the envelope at all.

New media, especially in the form of blogs and online newspapers, is becoming the medium of choice for journalists and citizens wishing to test official editorial boundaries. The internet was launched in Vietnam in 1997 but blogging did not begin in earnest until the mid-2000s when a crackdown on the mainstream media drew Vietnamese people unable to express themselves in the official press online, and the launch of Yahoo!'s Vietnamese-language 360° blogging platform made blogging easier. Most of Vietnam's political bloggers are urban professionals, often young and drawn from the ranks of journalism (although other professions like academia, law, and medicine have also produced many bloggers). They cover a range of politically sensitive issues, including corruption (particularly sensitive is discussion of high-level corruption), Sino-Vietnamese relations, territorial issues, policy and policy making, democracy, religion, and human rights. Online newspapers, or the online versions of more risqué print papers, are increasingly covering stories that have not been officially sanctioned in the knowledge that they can remove them from their website if instructed to do so…but only after stories have been picked up and shared around the blogosphere.

2 Public Demand for Coverage of Climate and the Environment

Vietnam is one of the countries to be most strongly impacted by climate change. Its 'acute vulnerability' results, according to Fortier (2010, p. 230), from 'high exposure to natural elements; high sensitivity of socioeconomic structures to such elements; and low capacity to adapt by protecting those structures or making them less sensitive.' Fortier notes that the country has a low-lying coastline of over 3,000 km with myriad deltas and floodplains, while at the same time it is susceptible to monsoons and typhoons. All of these factors make it 'widely exposed to sea-level rise…and weather extremes,' (Fortier 2010, p. 230). In inland areas, changes in rainfall and other weather patterns linked to climate change are resulting in droughts, landslides, changes to biodiversity, and reduced food productivity. As a result 'Vietnamese and international actors are now paying much attention to the issue of climate change,' (Fortier 2010, p. 231).

However, the official government response focuses largely on the assumption that Vietnam is a 'victim' of imported climate change, and that it should therefore be the recipient of aid in order to tackle the problem. While this is partly true, the belief—widely shared with the public through the media—that Vietnam can and should sit back and wait for help means policy makers are largely ignoring the role of home-grown environmental problems. It 'short circuits the understanding of the multiple ways in which human activities produce emissions and eventually lead to climate change…It de-socialises and de-politicises the problem, situating it in a physically-defined space of human-nature relationships, rather than in the more unsettling space of social relations' (Fortier 2010, pp. 237–238).

There is a 'very narrow Vietnamese climate change debate, apparently under heavy influence from key stakeholders…and with limited inputs from other stakeholders: vulnerable population groups, civil society, critical journalism, or critical academics,' (Bruun 2012, p. 15) and the information about climate change that citizens receive through the official media tends only to reflect this narrow debate.

Home-grown environmental damage is an equally serious issue, about which most people, especially those outside major urban centres, remain largely unaware or unconcerned. Vietnam is a fast developing nation with a rapidly growing population, and human pressures on land and other resources are having a significant impact on the environment. In Quang Nam, one of Vietnam's central provinces, Bruun found that 'in all crucial areas of resource use… including land, water and forests, environmental protection is wrought with problems and mostly losing out to economic exploitation. New economic opportunities are eagerly embraced and create wealth for some…but at the same time risking new collective and individual vulnerabilities,' (Bruun 2012, p. 15). Across the country, many of those who accumulate this new wealth are connected to the state and take advantage of their relationship with local seats of power to gain material wealth. Because Vietnam's media is owned by the state it struggles to uncover such corruption, meaning the disclosure of information to the public about the struggle between human wealth and environmental protection may be stifled.

Because few people are aware yet of the extent of the externally imposed and home-made environmental dangers facing Vietnam, there is limited demand for information about these topics. Indeed, the rush to modernise is pushing longer-term environmental issues from most people's minds, and this is especially true in poorer areas, which include much of the central region. While media discussion of climate change and environmental degradation is possible and does happen, it is not seen as a priority by most news outlets. "There's not much media coverage. It's an important issue, but it's not seen to be as important as politics, economics and other social issues," said a journalism professor teaching at the University of Hue. This is because most members of the public "don't care yet about environmental issues. They just want to focus on income generation."

This view may be changing: Hanoi's air is now among the world's most polluted, largely as a result of particulate matter released by traffic and construction sites. Ho Chi Minh City also ranks highly. With Vietnam's media based in these cities and increasingly trying to cater to public and not official demands (see below), coverage of issues such as air pollution are growing as citizens of these major cities demand change. Coverage by national media organizations will also reach audiences outside of these metropolitan areas, slowly helping to raise awareness of, and interest in, environmental issues throughout the country.

3 The Role Played by Media

Vietnam's media plays three distinct roles: It is a conduit for the dissemination of top-down information, an avenue for investigation and enquiry, and a forum through which different sectors of society can initiate debate. Because of its history and the political, economic and professional constraints media organizations operate under, the old mainstream media offers high-quality service to its audience in only the first of these three but is beginning to show signs of progress in the latter two. New media offers these two almost exclusively.

3.1 Information Dissemination

All official Vietnamese news media is owned by the state and has traditionally been a propaganda tool used to disseminate information to the masses. Despite a raft of other economic liberalizations as Vietnam reforms its economy, the current prime minister has made it clear that press privatisation will remain banned for the foreseeable future. The media is considered a strategic industry and the ruling Communist Party clearly intends to retain control over it. Media consumption in Vietnam is extremely high, thanks to high literacy rates, a history of Communist Party propaganda that helped build both the means of distributing a message effectively and the habit of absorbing it, a rapid increase in rural electrification over the past decade that has facilitated broadcast transmission to even remote areas, and a similarly rapid growth in the telecommunications industry that has bought mobile phones and the internet to city and rural dwellers alike.

The country produces hundreds of newspapers at all national and sub-national levels, as all state-owned entities are allowed to produce a newspaper or periodical if they chose to do so, although private ownership remains strictly banned. Quality varies dramatically, as does readership, but newspapers of some form or another are available for, and are read by, almost all levels of society in all geographic locations. Literacy in Vietnam exceeds 90 %,[1] making the written word widely accessible. A system of community notice boards, where major papers are posted each day, and a habit of sharing newspapers once they have been read, mean that even the poorest communities have some access to the print media.

Television and radio are also available nationwide, as over 90 % of the country is electrified and almost all areas can receive broadcast signals. The broadcast media retains tight links to the state, as discussed above, so tends to offer a less nuanced take on the news. But its reach and immediacy makes it ideal for spreading information quickly, and broadcasters are well used by the state when emergencies arise. Their effectiveness does depend, however, on an uninterrupted power supply, and power is often lost during climate-change related emergencies.

In central regions where natural disasters are relatively common and people must be quickly alerted if a storm, typhoon, or flood is coming, Vietnam's old-school linear information transmission systems work to tremendous effect. The VNA publishes a news alert that is instantly picked up by all major news outlets and broadcast on television. "Almost everyone watches TV, so it's a very direct line of communication" to affected communities, said the journalism professor in Hue. But in 2009, when Typhoon Ketsana hit central provinces with unexpected force and strength, it knocked out power in much of the region and left broadcasters unable to reach their intended audience with news of higher-than-anticipated floodwaters. According to JICA (2009, p. 9), the use of media to warn citizens in Quang Ngai province about the typhoon 'went well…However, the challenge start(ed) at the commune level, and bring(ing) the information to the ward and to the households. Due to power failure, usual communication systems did not work.'

In Quang Nam province, where Ketsana's floods and storms also knocked out power, 'battery radios were useful information sources for local residents' (JICA 2009, p. 7). According to the professor in Hue, people also used their mobile

[1] The CIA Factbook put it at 94 % in 2009, after Vietnam's most recent census, with male literacy at 96 % and female literacy at 92 %.

phones to access a dedicated weather channel targeting fishermen in the central region and others who need to be quickly alerted to changes in the weather. Mobile phone coverage in Vietnam is extensive and expanding fast. Evans (2010, p. 76) puts the number of mobile phones in the country in 2010 at 127 million, or 140 % of the country's population. It gives a conservative estimate of mobile phone penetration in 2020 at 151 % of the population and offers a 'high growth scenario' of 171 %. High growth can be expected: in 2000, the country had only 815,000 mobile phones, according to Evans.

The only chink in the armour of this top–down communication system is the fact that VNA, which launches the chain of information flows, may not receive information fast enough to get it to affected communities as the news agency relies on other government bodies that are continuously developing their skills and may not always provide information in a prompt enough manner, especially if operating under emergency conditions.

Because of its relatively low audience penetration compared to more traditional mainstream media forms, and because the online journalistic (blogging) community does not have the same access to official information as the state-owned media, new media plays only a limited role in the dissemination of emergency information to the public. Where it does disseminate, it offers information of a more long-term discursive nature that urban readers may take home and discuss with family members who are unlikely to access blogs themselves. "Although I don't expect farmers and workers to read blogs because they don't know how to use (or have access to) computers, (Vietnam's) population is very young and young people all over will begin to use the web. Then they'll start talking to their parents and grandparents," said one veteran blogger. Early in 2012 Dan Tri, an online newspaper, published a story about a study published by the Vietnam Environment Administration's Biodiversity Conservation Department and looking at the impact of economic development on biodiversity and the environment more generally in three central provinces: Quang Nam, Quang Tri and Thua Thien Hue. The story was picked up by a blog, baomoi.com, which further spread news of the issue.[2] Over time, the sharing of information via the internet about issues such as the long-term impact of development will filter through to rural audiences who now have only limited access to the web.

A TNS study (2009), found that 24 % of internet users surveyed in and around urban areas classify themselves as 'rural,' showing that web use is spreading beyond Vietnam's major cities. The survey also found that while 56 % of both urban and semi-urban respondents owned a computer, a substantial 39 % of rural dwellers also owned one.[3] Internet penetration is expected to reach almost 40 million people by 2020, up from around 24 million now (Budde 2010, p. 4), meaning that over time new media will increasingly reach people outside major urban centres and its role in information sharing may increase.

---

2 http://en.baomoi.com/Info/Growing-threat-to-biodiversity/6/221984.epi.

3 The figure is likely to be significantly lower for residents of truly rural areas away from urban centers, although no data was available at the time of writing to substantiate this assumption.

The government has made it clear that it intends to extend the internet's reach through the entire country (using communal Cultural Houses to provide access in poor rural communities where internet cafes and home computers are unlikely to penetrate for some time).[4] Thanks to rising prosperity, falling internet costs and better connection speeds, more ways to access the web (such as via mobile phones), and improving Information Technology skills[5] the number of people wanting and able to use the internet will grow in tandem. Vietnam now ranks among the developing world's top ten countries for Information and Communication Technology Development, according to measurements by the International Telecommunication Union, and a 2010 study by Yahoo!-TNS Media found that the amount of time spent online doubled from 2006 to 2008.[6]

With increased access to all forms of media, Vietnam's citizens will be exposed to more and more information about climate change and the environment. Government-led initiatives through schools and other social organizations and civil society groups are also raising overall awareness of some environmental issues.

3.2 Investigative Reporting and Awareness-Raising

As already noted, there is only limited space for debate in the official media and most media coverage and state-led awareness raising offers insight into only the official Communist Party line. This provides information about official policy, new legislation, conferences, and other properly sanctioned news items: "Most reporters still think like a propaganda tool. They report what they're told and there's very little real investigative journalism," noted a newspaper editor in Ho Chi Minh City.

Offering other viewpoints or questioning official policy about climate change and the environment is much more controversial than reproducing the party line, and few mainstream media outlets are willing or able to take the risk. This is especially because demand for such news remains relatively low due to much of the population being more concerned with short-term economic gain than longer-term environmental protection. Where it does happen, new media is taking the lead, although segments of the old media are also beginning to respond to market demand for discussion of topical issues—including climate change and environmental degradation—by running investigative stories and questioning government policy. Often old media does so when forced to follow the lead of online reporters.

Such coverage, however, is tightly aligned to audience demand and for now it is mostly the country's urbane media consumers in Hanoi and Ho Chi Minh who

[4] IT strategies developed in 2007 and planning to 2010 for Vietnam's northern, central and southern regions aimed 'to step by step apply information technology to commune-level agencies.'(Decisions 14/2007/QD-BBCVT to 16/2007/QD0BBCVT), Article 3a).

[5] Vietnam is developing its IT sector with a number of large foreign-invested projects, particularly in and around HCMC.

[6] The study found that Vietnamese people spend an average of 43 min/day online.

demand such politically sensitive reportage. In the two main cities "they have (television) channels discussing environmental issues, but here, because those channels are offered via cable, very few can afford to access them," said the professor in Hue. Localised, grassroots news organizations in other regions, including the centre, have little incentive to risk angering the authorities by questioning their actions, and most also lack the resources and training to do so, and so coverage remains focused by the large news organizations on issues of national rather than local importance.

But where an issue arises in the central region that has nationwide implications, coverage can be extensive. In the most notable case to date, a small group of academics were able to force a major change in government policy through the use of a single blog, http://www.bauxitevietnam.info/.[7] The blog was launched in 2009 by an academic named Nguyen Hue Chi to lobby the government to reconsider a $1.1-billion project awarded in large part to a Chinese company to mine bauxite in Vietnam's Central Highland region. Environmentalists believed the project would inflict serious environmental damage on the region as well as destroy local communities and give local jobs to imported Chinese labourers. The blog received over 17 million hits before being hacked in late 2009 and early 2010, after which it was moved to new sites (including http://bauvinal.info.free.fr/ and http://boxitvn.wordpress.com/) before finally succumbing to official pressures and closing altogether. Despite its closure, Bauxite Vietnam succeeded where no other media form had succeeded before: it forced the prime minister to return the project to parliament, where it was discussed for a second time and new environmental impact studies were demanded.

Blogs like the bauxite site "can have significant influence, especially when policy is being made. (National Assembly delegates) used to monitor the mass media to gauge public opinion. Now they monitor blogs," noted a senior Vietnamese journalist interviewed for this paper.

The site also exposed high-level intra-party discussions, forcing a degree of transparency not often seen in Vietnam. When a group of retired war veterans wrote to the Politburo demanding that officials tainted by the bauxite case be reprimanded or removed their letter was circulated through the blogosphere. Senior Politburo officials responded by telling the veterans to air their concerns in private, and rumours circulated that Politburo member Trung Tan San had visited one retired general and a tense conversation had ensued. The general responded to these rumours by posting a note on the bauxite blog explaining that the conversation had in fact been cordial. Interviewees said that by doing so he dispelled rumours (and thus pacified the Politburo) while also keeping the discussion in the public domain.

In this way, blogs are raising reportage of corruption in a way that the old mainstream media struggles to do. And as much of the corruption in Vietnam relates to the ownership and use of land, increased oversight in this area could significantly impact the way Vietnam views protection of the environment. Less widely read blogs can also impact decision making and government action at the national and lower levels, especially when bloggers force the mainstream media to run stories

[7] The site has had numerous web addresses as its creators have attempted to stay ahead of censors who have periodically blocked the site. This is its first web address.

that may otherwise have been ignored or spiked and, by doing so, increase public access to information about the issue under discussion. Such ‘follow up’ stories by the mainstream media are increasingly common as their urban readers write to demand follow-up coverage of an issue they have read about online.

When the National Assembly was asked by the government to debate a $56-billion high-speed rail link between Ha Noi and Ho Chi Minh City (HCMC) in May 2010 bloggers galvanised the media to cover the issue, raising questions about the project’s financial viability and the debts it would create for Vietnam. Assembly delegates later rejected the project. “Even if the media doesn’t pick up the story, the blogs will continue to cover it” and by keeping it in the public domain these bloggers raise awareness of the issue under discussion and “anger at the abuse grows,” said a blogger and journalist in Ha Noi. Eventually “someone with a motive will see it and force some kind of action…it’s infighting (amongst government officials) but it’s effective nonetheless.”

These blogs, although numbering only in the hundreds, are thought to have a significant impact. They influence the public both directly (because reading blogs can affect people’s knowledge and opinions) and indirectly, as media coverage and sometimes also government actions are influenced by bloggers. A few policy makers read and monitor blogs themselves, using blogs as a source of news and information about public opinion. But although direct readership by government officials is limited, interviewees said many officials are strongly influenced by blogs via the media, which they believe both reflects and forms public opinion.

There is “not enough evidence to judge” how many of Vietnam’s leaders personally read blogs, said a blogger in Ha Noi, although he and other interviewees believe several key leaders either monitor or instruct their secretarial staff to monitor blogs they think may be opinion-forming and/or critical of the government: “They don’t speak about it but they do read blogs. I have personally printed some out and given (copies) to them,” said a newspaper executive in HCMC. Leaders who do not directly monitor the blogosphere are made aware of the issues being discussed in it by family and friends. As one interviewee in HCMC pointed out, “their children read blogs and say ‘Daddy, they’re writing about you’.” Although “some don’t like blogs (they) want to know what people are thinking,” he added.

The online media already plays an important role in disseminating information to the mainstream media and forcing traditional outlets to cover stories and issues they may otherwise have ignored. Blogs are increasingly breaking news stories that members of the Vietnamese media were either unaware of or unable to publish. Bloggers sometimes also scan documents that they wish to put into the public domain and will either publish them on their blog or offer them to an overseas website. “One blogger exposed a big scandal about a senior leader of a newspaper” by posting secret documents online, a blogger in Ha Noi noted.[8]

These blogs are “a very important, helpful, and effective tool (for the media). They provide a very good source of information and ideas” and by doing so, open up discussion of issues that may otherwise remain taboo and force the media to

8 Further details of this case were not immediately available.

react, said a blogger and journalist in HCMC. "A lot of stories are now broken on blogs and the (mainstream) media has no choice but to follow up," she added.

Because bloggers must filter the information they post on their blog from a huge range of online news resources, journalists reading well-respected blogs tend to think of them as news gathering and filtering services and will scan them every day for news and story ideas. Some blogs, like Thong Tan Xa Vang Anh (http://ttxva.com/, or 'Vang Anh News Agency') go one step further, turning their blogs into 'citizen news agencies.' Popular 'web spider' blogs like http://anhbasam.com and http://anhbasg.multiply.com collate links to different news stories and sources and link readers to other blogs containing both republished official news stories and original content. "Blogs can link to good articles, including some that aren't published in the official media. Bloggers are intellectuals, they give reporters ideas and make them more articulate in their thinking," said a retired senior news editor in HCMC.

It is not only blogs that are influencing the Old Media. Economic reforms are significantly altering the way the media works and thinks, making editors more willing to challenge the official line and in the print sector particularly there have been a number of slow but steady moves away from state control. As part of a larger policy of offloading state assets, there has been a significant shift away from state funding of media operations to self-funding, and this has forced newspapers (and some broadcasters) to gain market savvy and focus as much on market demands as on political ones. In recent years a previously docile media has begun to question its former state paymasters and listen more intently to its newer market ones (both advertisers and audience) and this has made segments of the mainstream media more attune to audience needs and demands. While much of the market shows little interest in issues relating to climate change or environmental protection and thus coverage remains limited, one area of news coverage that has implications for these but is also of strong economic interest to readers is land.

As the government and major businesses requisition land for infrastructure and other projects, those they take land from are often forced to sell at significantly below market price: "With the signature of an official you can make one person very rich and push thousands of families into poverty," said a news editor in Hanoi.

Such land transfers have been taking place for over two decades, since the launch in 1986 of Vietnam's economic reform programme, and anger over these land grabs has risen to levels that the government can no longer ignore. The media is now allowed to cover corruption in the land sector so that policy makers can gauge the levels of discontent and work out how to address it. By doing so, news organizations not only offer useful information to policy makers but also give citizens an avenue through which to air their grievances. Over time, this is making people more willing to talk openly to members of the press (something that in the past few would do for fear of official reprimand), and is also raising awareness of some issues that may not otherwise have made it into the public sphere.

In central coastal areas around Danang, substantial tracts of land have been requisitioned to make way for hotels and other tourism developments, with implications for the local fishing and aquaculture industries and for coastal environments more generally (McKinley 2008). In the Mekong Delta, a policy decision to increase shrimp farming for export has lead to increased salinization of delta waters, meaning that farmers will be unable to return to rice cultivation if ever shrimp farming becomes unviable.[9] In the Central Highlands, a similar top-down policy decision to increase coffee cultivation in the 1990s forced traditional rotation crop farmers from their land. Because of excessive plantation development following a rush for wealth, forest lands were denuded and valuable top soil was lost. By reporting cases of land loss as these policies are put in place, Vietnam's media is beginning to both understand, and raise public awareness about, the broader environmental implications of rapid economic development.

Still, because of the resources needed to investigate such stories, coverage is largely the job of major newspaper groups, meaning that audiences in central Vietnam are only likely to be exposed to investigative reporting that tackles issues of direct concern to them if a story is picked up by a national news organization. This is happening increasingly often: In the spring of 2012, a number of national newspapers covered a story about cracks appearing in the Song Trang 2 Hydropower Plant dam in Quang Nam province, where strange earthquake activity was also thought to be linked to pressures within the power plant's reservoir when it was built. The story gained nationwide coverage, helping raise awareness around Vietnam of the dangers associated with the damming of rivers for power generation.

But there are limits to the degree of coverage the media can offer. Some are imposed by the state—where it believes coverage of a story would impinge on the interests of the state or of the Communist Party—but others come from within the industry itself.

3.3 Media Skills Deficits

A significant limitation stopping the Vietnamese media from facilitating awareness of environmental issues or asking people to question their assumptions is a shortage of reporting staff with the requisite investigative skills and sectoral knowledge. Investigative journalism differs from general news journalism as it requires greater investment of time, skills, and resources to cover a story and demands higher ethical standards. These qualities, where they exist, are contained within a small number of urban news organizations, particularly print and internet-based, that have already been discussed. "Big papers get everything they need. Smaller papers don't have the resources (to write investigative stories) and they dare not (do it) anyway," said one reporter.

Professionalism within the Vietnamese media is improving, but from a very low base. The introduction of competition into the media market has forced editors to

[9] The Mekong Delta is Vietnam's traditional rice basket, producing significantly more rice each year than either the centre or the north of the country.

fight for readers and viewers by raising content quality. "Compared to ten years ago we're getting better and better," said one editor. Yet even within the major newspapers, editorial staff continue to complain of low journalistic quality and inadequate opportunities to raise skill levels or use new skills once they have been earned.

Vietnamese journalists are hired mostly on graduation from journalism school where 'the curriculum is heavily oriented towards theory and the mode of delivery is predominantly lecture-based' (Rowland and Schofield, 2009, p. 2). This means that few new journalists are prepared for the everyday realities of their profession, and overall practitioner skills vary considerably, with 'pockets of excellence' (McKinley 2008, p. 10) interspersed among a larger population of mediocre-to-poor-quality journalists.

Because most high-quality journalists work in Hanoi or HCMC for national news organisations or sub-national ones with a national audience, smaller news organisations find it difficult to find and retain good reporters and editors. Universities in both Hue and Danang run journalism and communications programmes and graduate around 150 students a year between them, but these graduates are mostly unable to find work in news organizations with the capacity to further their professional training, thus professionalism stultifies in Central Vietnam. There are very few opportunities for journalists and editors to improve their skills. "We really need training. We can't travel to the big cities because it's too far and too expensive, and they won't take us (on training courses) because we don't have enough experience," said a print editor in Hue.

The weaknesses inherent in much of the Vietnamese media, particularly outside these two major centres, mean that 'certain parts of the press have limited credibility among an increasingly sophisticated and media aware audience' (Rowland and Schofield 2009, p. 1). Editorial abuses as well as errors are common. Many journalists "consider themselves very well educated," and allow themselves "too much power" when writing stories, according to one editor in Hanoi. "When they lack facts they end up writing their own judgements…it's lazy but it's easy," he said.

The concentration of investigative reporting power and skills within a small number of news organizations could facilitate the creation of an elite group of journalists and editors who, over time, will gain significant power and become increasingly willing and able to confront established editorial norms. Over time, they might offer 'trickle down' training and standard setting to reporters in smaller news organizations who will then use their skills at the local level. Alternatively, the polarization of reporting power may draw resources away from smaller community news organizations that are geographically well placed to cover environmental stories at society's grass roots but employ reporters who are woefully under-equipped to do so. Journalists working on investigative stories need substantial resources in addition to strong investigative skills: time and money are key as are more day-to-day resources like internet access, a working tape recorder, and desk space. Only Vietnam's wealthiest news organizations can attempt to offer these things.

Limited access to the resources needed for investigative reporting (not least, information, but also time, money, and sometimes technical resources) can seriously hamper journalism quality and professionalism. For smaller, local news organizations it can make coverage of things like natural disasters extremely difficult, if staff lack basic tools such as waterproof cameras in a storm or flood zone.

4 Providing a Forum for Discussion

The public increasingly view the media as a route through which to make their discontent at land grabs, unequal development, and other issues felt by policy makers. Letter and comment pages in newspapers, call-in radio programs, and internet blogs all facilitate upward communication and allow Vietnam's often voiceless poor to air their views. In order to retain social stability, the government must hear these voices so that it can tailor policies accordingly, and so far it seems willing to countenance the development of audience feedback through the mass media.

Public comments are becoming a major source of information for journalists and one of the main routes through which they identify and investigate stories that might otherwise never see the light of day: "It's often letters of complaint sent to the paper by people who are unhappy with judicial hearings or who feel they have been treated unfairly" that offer early insight into a story, said a senior editor in Buon Ma Thuot.

The public are becoming more willing to contribute to general news stories as well. One reporter in the central highlands described how she researches her stories: "Accessing information is never easy, especially when you're researching a 'negative' story. First, you have to study as much written evidence as you can, then go on a fact-finding mission to gather as much anecdotal evidence as possible before attempting to interview an official. Disadvantaged people are usually very free when you talk to them, but the reverse is true of local officials."

Freedom of public expression through media forums, especially online, is growing and there are signs that the government understands a growing public need to use the media as a space for debate. While the public's voice is increasingly being heard over the airwaves, on the internet, and on the letters pages of newspapers, civil society's timid declarations are also slowly gaining in tenor. Like media, there is no truly independent civil society in Vietnam. Non-governmental voices that don't echo the official line are seldom heard through the state-owned media, and because Vietnam's civil society is small and underdeveloped, very few non-governmental organizations have learned how to use new media effectively (although increasingly their activities are being covered by the media).

Groups commonly called non-governmental organizations (NGOs) must in fact be organised under the Fatherland Front, an umbrella organization belonging to the state. A Law on Associations—long discussed but not yet passed by parliament—might improve the plight of civil society groups by allowing them a degree of independence, but this legislation, if it comes, will be some way off. Most of Vietnam's civil society comes in the form of 'mass organizations', groups operating also under the Fatherland Front and designed to represent the interests of various sectors of society, such as farmers, workers, and women. They work primarily on service delivery, not advocacy, but a small number are beginning to combine the two roles and are offering citizens a route through which to communicate with government.

The Social Policy Ecology Research Institute (SPERI), for example, 'explains to local and national governments the value of local knowledge about farming methods, forestry management, biodiversity conservation…and the concerns various communities have about commercialised farming…and demonstrate the important roles NGOs can play in conveying citizens' voices, in particular poor people's voices,'(Kerkvliet 2009, pp. 35, 38).It undertakes this role as a result of 'experience in other activities, especially delivering services,' Kerkvliet adds.

Although there is as yet limited cooperation between the media and these civil society groups, there are signs that it is beginning to grow. The Vietnam Forum for Environmental Journalists was born over a decade ago from a marriage between the Vietnam Journalism Association and the Vietnam Association for the Conservation of Nature and the Environment: one media body and one environmental NGO. It operates an educational website for reporters covering environmental issues and, when funds are provided by international donors, offers training courses as well. It hopes to affiliate with groups of environmental journalists overseas in order to learn from them and share ideas, according to its Executive Deputy President and day-to-day manager, Huang Quoc Dzung.

In his conclusion, Kerkvliet finds that Western media and academia 'convey a picture that Vietnam remains a largely closed political system in which nearly everything is under the control of the state. However, there is evidence of a developing and widening civic space for ordinary Vietnamese to pursue common interests (and)…that many officials and agencies are willing to engage these citizen groups.' The financial reforms now taking place within the media will, over time, increase journalists' access to the skills and resources they need to develop this space. Cooperation with a slowly growing civic sector will also facilitate change. The media's ability to inform the public about climate change and environmental issues, and to offer them a forum through which to discuss these issues is growing. The change is slow but the trend toward a more open media space is clear.

References

Bruun O (2012) Sending the right bill to the right people: climate change, environmental degradation and social vulnerabilities in Central Vietnam. Weather, Clim Soc 4(4):250–262

Budde P (2010) Vietnam—internet market. Paul Budde Commun

Evans P (2010) 'Vietnam: telecoms, mobile, broadband and forecasts. Paul Budde Commun

Fortier F (2010) Taking a climate chance: a procedural critique of vietnam's climate change strategy. Asia Pac Viewpoint 3:229–247

JICA (2009) Assessment report on response to typhoon ketsana in the central region. Project report for building disaster resilient societies in central regions of Vietnam

Kerkvliet B (2009) Forms of engagement between state agencies and Civil Society Organizations in Vietnam. NGO Resource Centre, Department for International Development, and Embassy of Finland in Hanoi

McKinley C (2008) A study of the Vietnamese Press coverage of land issues as they relate to poverty reduction. Ford Foundation, New York

Rowland B, Schofield B (2009) MediaPro project: interim report

TNS Media Vietnam (2009) Media habits survey: wave 2, 2009. Power point presentation

Yahoo!-TNS Media (2010) Digital Vietnam: internet media habits decoded (net index 2010 highlights). Power point presentation

Health Impacts of Climate and Environmental Change: Awareness and Challenges to Adaptation

Peter Furu and Duong Khanh Van

Abstract Taking a point of departure in current knowledge about the existing health situation in Vietnam and insights into interrelationships between social, environmental and institutional risk factors and health, this chapter reports on small-scale health surveys conducted in purposely selected communities of the Thang Binh district of Quang Nam province. The surveys aimed at exploring awareness of and coping to environmental and climate change induced health problems. In the surveys most respondents associated climate change with abnormal weather conditions and typically mentioned seawater level rise, storms, floods and increase in temperature. Generally, respondents had observed considerable changes in health patterns in recent years however, without linking these clearly to climate change or climate factors but rather to a change in environmental determinants of health such as food, water and air quality. The observations are consistent with a wider socio-economic household survey, reporting a worsening of the health condition by a majority of households. A range of coping mechanisms were highlighted by community members and local authorities as ways and means of protecting each other and individual households and their members in times of increased extreme weather events and general environmental change. Future new actions should ideally be informed by parallel research initiatives and the present small-scale survey may stimulate more in-depth and broader studies that may help to identify proper, sustainable solutions for future adaptation and coping to climate and environmental change.

Keywords Health impact of climate change • Perceived impact • Fatalities • Vulnerability

P. Furu (✉)
Department of International Health, Immunology and Microbiology, University of Copenhagen, Copenhagen, Denmark
e-mail: furu@sund.ku.dk

D. K. Van
National Institute of Occupational and Environmental Health, Ministry of Health, Hanoi, Vietnam

O. Bruun and T. Casse (eds.), *On the Frontiers of Climate and Environmental Change*, Environmental Science and Engineering, DOI: 10.1007/978-3-642-35804-3_11,

1 Introduction

This chapter will highlight some of the overall global relationships between development, environment, climate and health as a basis for understanding the present health situation in Vietnam in general and in Quang Nam province in particular. The Government of Vietnam is much aware of the challenges facing society from the combined forces of anthropogenic and development-induced climate and environmental change. This is for example reflected in the *National Target Program to Respond to Climate Change* (NTPRCC) (GOVN 2008a) approved by the Vietnam Government in 2008 (GOVN 2008b) as well as in the National Climate Change Strategy (GOVN 2011).

Table 1 illustrates the level of sector-specific climate change impacts as estimated by the Government in connection with its response planning. However, the NTPRCC recognizes that there is a need for continuously establishing new evidence on the extent of climate change and its impacts on every sector, area and locality (GOVN 2008b). It is envisaged that the health sector will experience medium-to-high impacts, demanding concerted efforts in addressing the health challenges that particularly vulnerable population groups will be confronted with. As a consequence, the Ministry of Health (MOH) has agreed on an action plan specifically to protect community health from the adverse impacts of global climate change (GOVN 2010).

As a contribution to increasing the knowledge base for climate change related decision-making, we conducted preliminary surveys on awareness and coping mechanism in relation to climate change impacts on health in selected communities and local authorities in Quang Nam province. The results should be seen as an encouragement for further studies on the complex interactions between environmental, social and institutional determinants of health.

Table 1 Potential Impacts of climate change and sea level rise

Sector, area, object	Impact factors					
	Temperature rise	Sea level rise	Tropical cyclones	Flood	Drought	Other climatic extremes
Agriculture and food security	High	High	High	High	High	High
Aquaculture	High	High	High	High	Medium	Medium
Energy	High	Medium	Medium	Medium	High	Medium
Industry	High	High	Medium	Medium	Medium	Medium
Transportation	High	High	High	High	Medium	Medium
Construction	High	High	High	High	Medium	High
Tourism	Medium	High	High	High	Medium	Medium
Health care	High	Medium	Medium	High	High	High
Natural ecosystems and biodiversity	High	High	Medium	Medium	Medium	Medium
Water resources	High	High	Medium	High	High	Medium
Residential area	Medium	High	High	High	Medium	Medium

Source GOVN (2008a)

2 Global Development, Environment, Climate and Health Inter-linkages

Overall interfaces between determinants of health are generally elucidated as part of epidemiological investigations and development studies. The World Health Organization (WHO 1997) has established a model (the so-called DPSEEA model) explaining the causal relationships between development, environment and health. In this model the global earth and human systems are considered constantly under pressure by the population growth and economic development, which is experienced in most parts of the world (including Vietnam). Production and manufacturing systems as well as consumption patterns and the resulting waste released, exert a tremendous pressure on global climate and environment. As a consequence, the state of the environment is characterized by interacting natural and man-made hazards, resource scarcity and high pollution levels.

Depending on the level of human exposure to, and resilience to, existing and new hazards, health effects may be seen in terms of changes in well-being, disease and deaths. The DPSEEA model as shown in Fig. 1 illustrates this cause-and-effect relationship between development, environment and health. Various adaptation and mitigation actions may be targeted at various entry points in this hierarchy of interrelatedness (WHO 1997).

Climate change is only one of several human-induced environmental changes caused by the fast increasing human pressure on the global and local environments. Transformations include land cover changes (including forests), land degradation, desertification, wetlands loss, biodiversity loss and others as established by the Millennium Ecosystem Assessment (2005) and illustrated in Fig. 2.

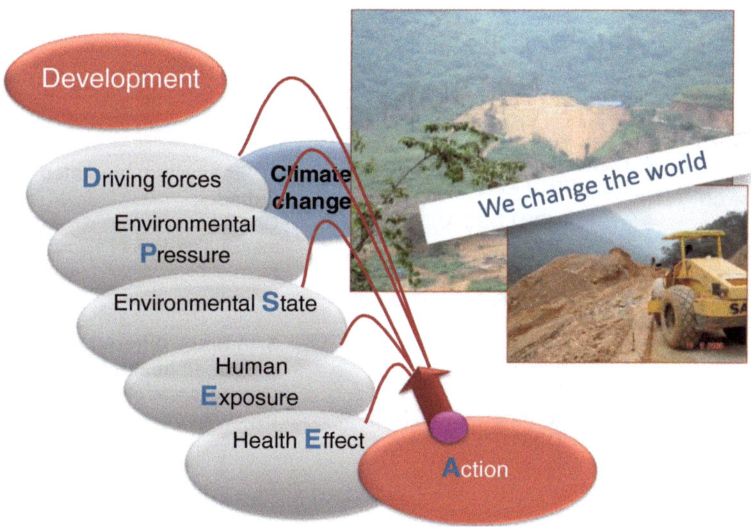

Fig. 1 The DPSSEA model (modified from WHO 1997)

It is important to note that health is influenced not only by climate change, but also by common weather and climate variability as well as by overall social and environmental change. Identifying the causal relationships furthermore remains a challenge because most human health conditions are multi-factorial, and the underlying socioeconomic, demographic, and environmental context changes significantly over time (IPCC 2001). In particular, the local development-related changes may confound the changes seen from climate change. Similarly, economic globalisation represents a double exposure in synergy with climate change (O'Brien and Leichenko 2000).

2.1 Types of Climate Change Related Health Impacts

When attempting to establish the linkages between health and climate related factors some definitions are considered important. *Climate* is the average state of the atmosphere and the underlying land or water in a particular region over a specific time period; *weather* is the day-to-day manifestation of climate in a particular place at a particular time; *climate change* is a statistically significant variation in either the mean state of the climate or in its variability persisting over an extended period (typically decades or longer); and *climate variability* refers to variations around the mean state, including the occurrence of extreme weather events (Kovats et al. 2003). With reference to the Intergovernmental Panel on Climate

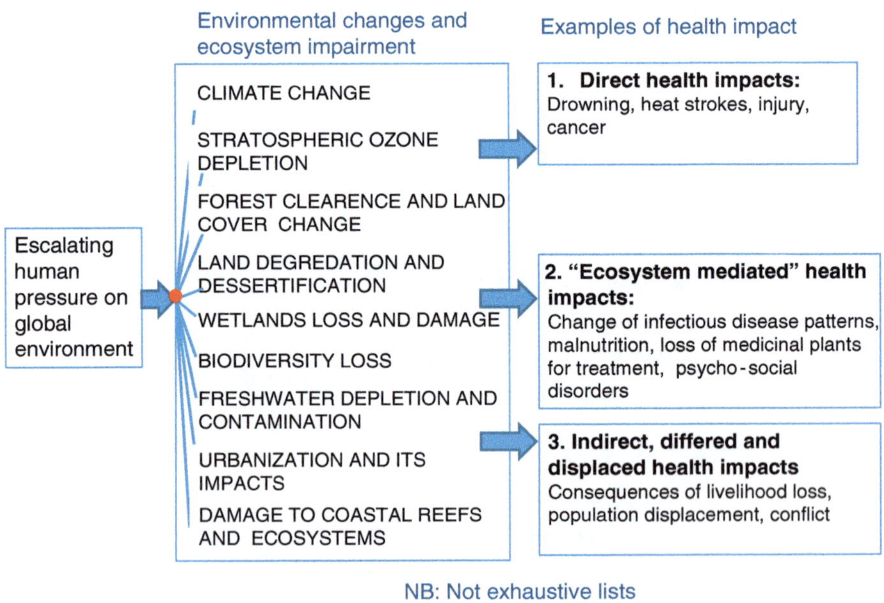

Fig. 2 Examples of health impacts of environmental change (modified from Millennium Ecosystem Assessment 2005)

Change (IPCC 2001) WHO states that "*broadly, a change in climatic conditions can have three kinds of health impacts:*

- *Those that are relatively direct, usually caused by weather extremes.*
- *The health consequences of various processes of environmental change and ecological disruption that occur in response to climate change.*
- *The diverse health consequences—traumatic, infectious, nutritional, psychological and other—that occur in demoralized and displaced populations in the wake of climate-induced economic dislocation, environmental decline, and conflict situations*"(WHO 2003).

Peoples' livelihoods, community vulnerability and the risk of suffering the above three types of health impacts are likely to be determined by a number of factors linked to climate-related environmental conditions. These include changes of for example temperature, water availability (floods, droughts), breeding potential for vector mosquitoes, soil condition and its suitability for agricultural productions, as well as the capacity of the health systems to cope with prevailing health conditions at local levels. Thus, vulnerability is considered partly a measure of the functionality of programs and instruments in place to reduce the burden of disease, and partly a measure of the success of traditional public health functions, such as providing safe water and sanitation and responding to outbreaks of infectious diseases (e.g. malaria). Furthermore, general poor accessibility to primary health care services contributes to high levels of vulnerability and low adaptive capacity of millions of people (Confalonieri et al. 2007).

Some of the considered health effects of climate change in Southeast Asia include morbidity and mortality due to heat stress; vector-borne infectious diseases (for example malaria and dengue); diarrheal diseases; and malnutrition. The effects of extreme weather events such as cyclones, flooding and landslides after heavy rainfall may include injuries and deaths. Respiratory tract diseases caused by increased air pollution and psycho-social disorders related to social dislocation and migration could be attributed indirectly to climate change (ADB 2009).

3 National Health Trends in Vietnam

The health status of the Vietnamese population has changed dramatically in recent years, primarily as a result of the economic reform or renovation program known as *Doi Moi*, introduced in 1986 (see "Paradoxes in Adaptation: Economic Growth And Socio-Economic Differentiation. A case Study of Mid-Central Vietnam"). This led to a considerable poverty reduction including improved livelihoods, better living conditions and better environmental and health services (GOVN 2006; WHO 2011a). According to WHO, the maternal mortality ratio decreased significantly from 130/100,000 live births in 1990 to 69/100,000 live births in 2009. Also the under-five mortality rate has fallen dramatically from 55.4 % in 1990 to 24.1 % in 2008 (WHO 2011a). Other selected key health indicators are presented in Table 2. Vietnam has witnessed an average growth rate of approximately 7 % over the last decade, which is reflected in changing demographic patterns, declining fertility declining, and a gradually increasing share of elderly people.

However, the above positive changes in health indicators are accompanied by negative changes in disease patterns of a range of health conditions. Thus, like many other low and middle-income countries, Vietnam is undergoing an epidemiological transition from the dominance of communicable diseases (CDs) to the increased prevalence of chronic, non-communicable diseases (NCDs) (Minh et al. 2009; Ha and Chisholm 2011). Over the last three decades, NCDs have shown a consistent trend of increase to become dominant against CDs and injuries, seen in Table 3. This is comparable to the overall trend in the Southeast Asian region, in which mortality from NCDs is expected to further increase by 21 % over the next 10 years (Narain et al. 2011).

This dominance of NCDs is reflected in Table 4, showing the ten leading causes of death in Vietnam in 2008.

3.1 Communicable Diseases

For some selected health issues in this category of diseases the continuous efforts by government and private sector health institutions have witnessed good reductions in morbidity and mortality with diarrhea and dengue fever as exceptions.

HIV/AIDS

Since HIV was first discovered in Vietnam in 1990 transmission has remained high in three high-risk groups: men who inject drugs, men who have sex with men and female sex workers. Since records began in the early 1990s, 249,660 cases have been reported with 52,325 AIDS-related deaths and 197,335 people presently living

Table 2 Selected national health and health risk indicators, Vietnam

		1990	2000	2009
Life expectancy at birth (years)	Males	63	68	70
	Females	67	72	74
Total fertility rate		3.7	2.3	2.0
Maternal mortality ratio/per 100.000 live births		170	91	56
Infant mortality rate (%)	Males	39	23	19
	Females	40	24	20
Under-five mortality rate (%)	Males	58	31	25
	Females	53	28	23
	Urban		Rural	
	1990	2008	1990	2008
Access to improved water source	88	99	51	92
Access to improved sanitation	61	94	29	67

Source WHO (2011b)

Table 3 Mortality in percent of total deaths (*Source* GOVN 2006; [a]) data from verbal autopsy (*Source* MOH 2010)

	1976	1986	1996	2003	2004	2008[a]
Communicable diseases	55.51	59.10	37.63	27.44	26.13	12
Non-communicable diseases	42.65	39.05	50.02	60.61	60.80	75
Injuries and accidents	1.84	1.85	12.35	11.95	13.00	13

Table 4 Top 10 causes of death in males and females in Vietnam in 2008

	Male			Female		
Rank	Disease category	Deaths	%	Disease category	Deaths	%
1	Stroke	53,217	18	Stroke	56,771	23
2	Liver cancer	19,915	7	COPD	14,941	6
3	Road traffic accidents	17,330	6	Pneumonia	11,175	4
4	Lung cancer	15,720	5	Ischemic heart disease	11,015	4
5	COPD	14,355	5	Diabetes	9,858	4
6	Ischemic heart disease	13,504	5	Liver cancer	8,587	3
7	Tuberculosis	11,450	4	Lung cancers	7,869	3
8	Pneumonia	9,470	3	Tuberculosis	6,798	3
9	HIV/AIDS	9,417	3	Road traffic accidents	5,750	2
10	Stomach cancer	8,469	3	Stomach cancer	5,470	2
Total		290,624			250,605	

Source Nhung et al. (2011)

with HIV/AIDS. The number of new HIV cases decreased between 2007 and 2009 and leveled out at about 14,000 new cases in 2010 and in 2011. HIV prevalence among the adult population (age 15–49) was 0.45 % in 2011 (NCADPPC 2012).

Tuberculosis

From 2005 to 2009 a small decrease has been observed in number of TB cases, from 95,970 reported cases in 2005 to 95,036 cases in 2009. The number of TB related deaths reportedly decreased from 1,936 in 2005 to 1,689 in 2009 (WHO 2011).

Diarrhea

Diarrhea remains one of the leading causes of morbidity in the country, particular in areas with inadequate access to safe water and appropriate sanitation. Table 2 points to the fact that in spite of progress in recent years, especially in rural areas sanitation coverage is still a challenge since more than 30 % of the population are without proper sanitary conditions. Cholera, typhoid fever and dysentery are the main causes of diarrhea. In 2008 there were 853 positive cases of cholera recorded in 22 provinces/cities, which was a 55.3 % reduction from 2007 (WHO 2010).

Dengue

The mosquito-borne dengue and dengue hemorrhagic fevers are considered large and increasing public health problems in Vietnam, with 128,831 cases of dengue in 2010 and 109 related deaths (MOH 2010b; WHO 2011a). There are currently indications that a potential geographic expansion in dengue activity is taking place from southern Vietnam, where 70 % of cases are reported, to central and highland areas. Within the region Vietnam is among the five countries with the highest dengue-related burden of disease.

Malaria

The national health authorities have ambitious plans to eradicate malaria in the country by 2015, reflecting that malaria morbidity continues to decrease: from 19,497 reported cases in 2005 to 17,515 in 2010. In the same period, however, mortality increased slightly from 18 cases in 2005 to 21 cases in 2010 (WHO 2011a).

Avian influenza

Avian influenza virus A (H5N1) has been recorded in poultry populations in Vietnam since 2003. Most human infections are linked to close contact to infected poultry. As per mid-2012 there have been 123 laboratory confirmed human cases of H5N1 with 61 fatalities (WHO 2012). Table 5 shows reporting to WHO on human cases of avian influenza by Vietnam and neighboring countries.

3.2 Non-Communicable Diseases/Conditions

As reflected in Table 2, Non-communicable Diseases (NCDs) dominate the current disease patterns. An increase in NCDs is seen in recent decennia with total morbidity rising from 39.0 % in 1986 to 66.32 % in 2009, and in mortality from 41.1 to 63.32 % (WHO 2011a). The leading causes for the epidemiological shift are the general economic growth and related life style changes as well as the fact that life expectancy has increased considerably. The main diseases are cardiovascular disease, diabetes, cancer, and asthma.

Cardiovascular disease

As seen in Table 4 stroke is the leading cause of death in both males and females, and ischemic heart disease is an important contributor to cardiovascular disease in both sexes. Furthermore CVD accounts for 40 % of total deaths for all age groups (WHO 2011c).

Diabetes

The obesity risk factor for diabetes is increasing throughout the country. In 2002, 2.7 % of the Vietnam population were estimated to have diabetes, of which 90 % were

Table 5 Cumulative number of laboratory confirmed human cases (C) and deaths (D) of Avian Influenza A (H5N1) reported to WHO

Country	2003		2004		2005		2006		2007		2008		2009		2010		2011		2012		Total	
	C	D	C	D	C	D	C	D	C	D	C	D	C	D	C	D	C	D	C	D	C	D
Cambodia	0	0	0	0	4	4	2	2	1	1	1	0	1	0	1	1	8	8	3	3	21	19
China	1	1	0	0	8	5	13	8	5	3	4	4	7	4	2	1	1	1	2	1	43	28
Lao PDR	0	0	0	0	0	0	0	0	2	2	0	0	0	0	0	0	0	0	0	0	2	2
Vietnam	3	3	29	20	61	19	0	0	8	5	6	5	5	5	7	2	0	0	4	2	123	61

Source WHO (2012)

type 2 (Baumann et al. 2006), and with an expected increase of diabetes type 2 to 3.5 % by 2025 (Beran et al. 2009). Overall the burden of disease with diabetes is highest in females, but peaks in the 60–69 age group for both sexes (Nhung et al. 2011).

Cancer

The cancer incidence rate is currently increasing with more than 100,000 new cases per year (WHO 2011a). In a national survey covering 2005–2006, liver cancer was ranked as most common (31,04 % for males and 19.91 %for females), followed by lung cancer (26.69 % for males and 18.21 % for females) and stomach cancer (14.42 % for males and 14.26 %for females). Among females, cervix and other female genital cancers had a prevalence of 9.13 % and breast cancer 5.69 % (Ngoan et al. 2007).

Injuries

Morbidity related to accidents, injuries and poisonings has seen an increase in the period 2002–2009, rising from 9.2 to 10.8 % of all hospital admissions. Similarly, mortality related to the above has increased from 18.5 to 22.6 % of all deaths in hospitals. Transport accidents represent the ninth leading cause of morbidity (WHO 2010) and the sixth leading cause of mortality (WHO 2011a). Drowning accounts for 17 % of total injury mortality and second to traffic accidents. Children is a particular vulnerable group with respect to drowning, with an average 22/100,000 children (age 0–4) drowning per year (Lan et al. 2011).

Malnutrition

According to WHO (2011a) protein energy malnutrition and micronutrient deficiencies among under-five children have seen a significant decline recently. Thus, malnutrition in the age group under-5 years was reduced from 33.8 to 18.9 % from 2000 to 2008. However, in the urban context and in more economically developed rural areas overweight and obesity are increasing in children with the potential, if uncontrolled, to develop into diabetes and cardiovascular diseases.

Table 6 Prevalence of 10 common mental disorders in Vietnam

Mental disorder	Prev. (%)
Alcohol abuse	5.3
Depression	2.8
Anxiety	2.6
Juvenile behavioural disorder	0.9
Old age amnesia	0.9
Slow mental development	0.6
Cerebro cranial trauma	0.5
Schizophrenia	0.5
Epilepsy	0.3
Drug use	0.3

Source Vuong et al. (2011)

Mental health

Although Vietnam has witnessed considerable economic and social progress, resulting in improvement in the general health status, the fast and sometimes uncontrolled development may lead to a series of negative impacts on the mental health status. Similarly, the increasing frequency of extreme weather events and resulting damage to infrastructure, housing, personal property and individuals may be conducive to a series of psycho-social disorders, such as post-traumatic stress.

Some of the key mental health issues prevalent in Vietnam are shown in Table 6 (please note that in a Vietnamese context these are identified as psycho-social aberrations).

Berry et al. (2010) suggest three pathways for mental health impacts of adverse weather events. First, mental health may be affected directly by the effects of more frequent and serious natural disasters on human settlements which, typically, cause serious anxiety-related conditions, and later, severe and chronic mental health problems. Next, extreme weather events may increase the risk of injuries and associated physical health problems, which eventually may lead to mental health problems. Third, climate change related events may change the natural and social environment on which people depend for their livelihoods and wellbeing, thereby potentially leading to mental health problems in vulnerable community groups. In a study on the social dimensions of vulnerability in Vietnam by Few and Tran (2010), increased anxiety from flooding-related hazards and resulting impacts on livelihoods was reported at household levels.

According to Vuong et al. (2011) about 12 million people are currently in need of mental health services. A nationally representative epidemiological survey on 10 common mental disorders in the period 2001–2003 showed that the 10 most common mental disorders had a combined prevalence of approximately 14.9 % of the population, corresponding to the above (see also "Rural Households: Socio-Economic Characteristics, Community Organizing and Adaptation Abilities").

4 Main Health Issues Relevant to Climate and Environmental Change

Vietnam is ranked as one of the countries in South East Asia particularly vulnerable to climate related hazards such as tropical cyclones, floods, landslides, droughts and sea level rise (Yusuf and Francisco 2009; ISPONRE 2009, see also "Impediments to Climate-Induced Disaster Management: Evidence from Quang Nam, Central Vietnam"). With a climate change related sea level rise of 1 m, and no adaptation measures, Vietnam would lose 28 % of its wetlands and over 10 % of its urban areas would be affected (World Bank 2008), inevitably having health implications.

Figure 3 represents the trend of impacts of natural disasters experienced in Viet Nam in the past decade. Millions of people are affected and the society suffers tremendous losses of lives and income generation. For example in the period

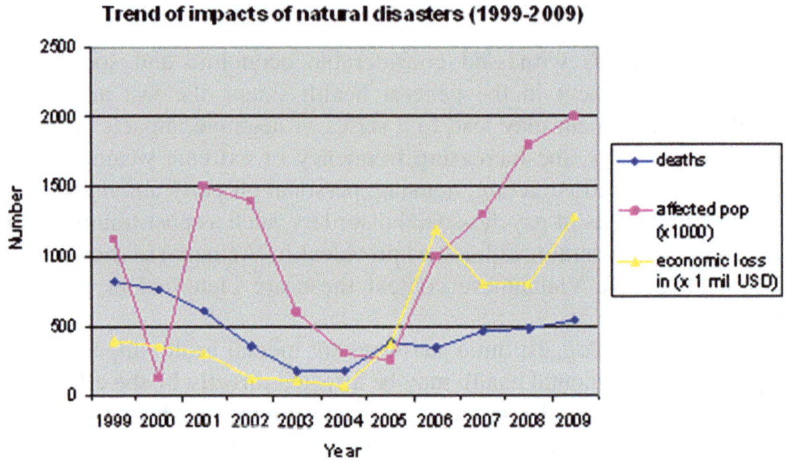

Fig. 3 Trends of impacts of natural disasters, 1999–2009, Vietnam (Binh et al. 2011)

2002–2006 1,700 people died in natural disasters and economic loss were estimated to VND 75,000 billion (GOVN 2007). As seen in "Impediments to Climate-Induced Disaster Management: Evidence from Quang Nam, Central Vietnam", Quang Nam ranks as one of the most vulnerable provinces to natural disasters, with more than half of the fatalities related to flooding events.

The broader potentially adverse impacts of climate change include effects on ecosystems, socio-economic conditions and human health, which are superimposed on the already serious effects of the overall development process on the natural and social environments. Based on comparative risk assessment, evidence synthesis and expert evaluations for regional exposure as well as on WHO country health statistics for 2004, WHO (2009) has estimated that about 21 % of the total burden of disease in Vietnam is related to environmental risk factors, equal to 122,500 deaths per year.

Diseases and health conditions relevant to climatic factors and climate change are those with climate sensitive determinants of health (both risk factors and health opportunities), i.e. where transmission would be related to climate and weather factors (e.g. precipitation, temperature, wind). Using time series of climate and disease variables and applying statistical and GIS spatial analyses, Kien et al. (2010) write in a recent study on linkages between climate change and variability and rates and distribution of certain infectious diseases in Vietnam, that there is good epidemiological evidence of associations between climate factors such as temperature, rainfall and climate variability on the one hand, and burden of disease of major infectious diseases such as malaria, diarrheal diseases and dysentery on the other hand. There were clear linkages between climate factors and the changes in disease patterns over the last 25 years, although for other communicable diseases like influenza, cholera and typhoid correlations were less obvious than those above. Additionally and indirectly, however, socio-economic factors sensitive to climate change and climate variability may also influence vulnerability, disease prevalence and distribution (e.g. income, accessibility to health services).

Vulnerable community groups

According to a review by Ahn (2009), with the existing pattern of gender discrimination in Vietnam women are more likely than men to die from the direct and indirect results of natural disasters. Women and girls also suffer more health impacts during, and in the aftermath of, natural disasters. For example, there are more drowning incidents among women because of lack of knowledge on how to swim (Vu Minh Hai 2004 cited in Ahn 2009). Furthermore, pregnant women and small children are particularly vulnerable to water-borne diseases such as diarrhea and cholera. As another vulnerable group, the elderly and people with poor health are at risk from climate change impacts like heat stress and malnutrition (Ahn 2009). People with disabilities are also vulnerable because of poor access to e.g. disaster risk reduction measures like preparedness training and early warning signals (Jørgensen 2011).

5 Awareness of Climate and Environmental Change in Thang Binh District, Quang Nam Province

Among its objectives for proper responses to climate change the Government of Vietnam has emphasized 'the need to promote scientific and technological activities to establish scientific and practical foundation for climate change responding measures' (GOVN 2008a, b). It is within this context that a small-scale survey was initiated in Quang Nam Province in order to establish the perceptions of climate and environmental sensitive health issues. In consultation with the Quang Nam Preventive Medicine Centre (PMC), a number of communities were selected in Thang Binh, a coastal district already prone to extreme weather events and flooding and vulnerable to climate change impacts.

5.1 The Survey Area

Thang Binh is a district in the eastern part of the Quang Nam province. The area is 3,847,507 ha with a population of 186,964 people (2006). The district has 21 communes and the District capital is Ha Lam. Thang Binh is bordered on the east by the China Sea, to the west by Hiep Duc District and Que Son District, to the south by Tam Ky town, and to the north Que Son District and Duy Xuyen District. The district has a tropical monsoon climate, with average annual humidity above 80 % and average annual rainfall reaching 2,000 mm. The climate here is divided into two distinct seasons of the year. The rainy season begins in September and lasts to February. The dry season begins in February and ends in August with hot sunshine, low humidity and generally dry conditions.

In Thang Binh land is divided into different regions, with mountainous and hilly areas occupying 2/5 of the district's land area: the coastal area is mainly sandy soil and the mountainous areas consist of dense forests and arid lands. Binh

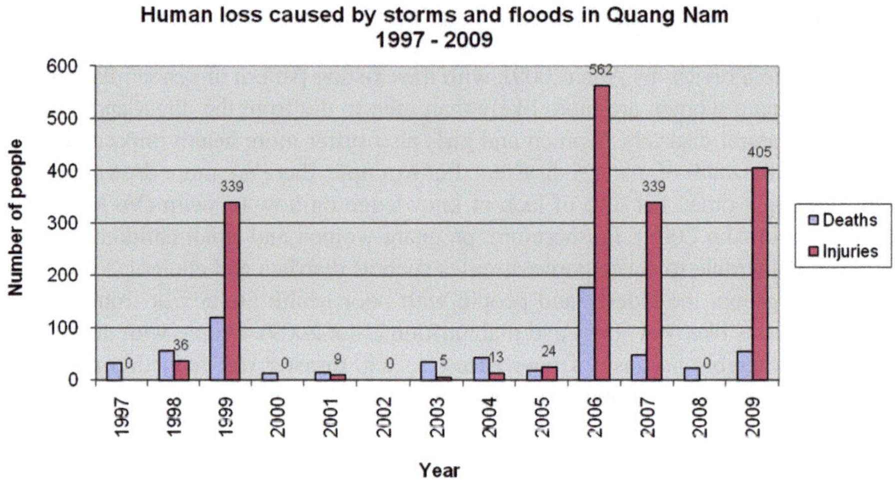

Fig. 4 Human loss caused by storms and floods in Quang Nam 1997–2009 (Bình et al. 2011)

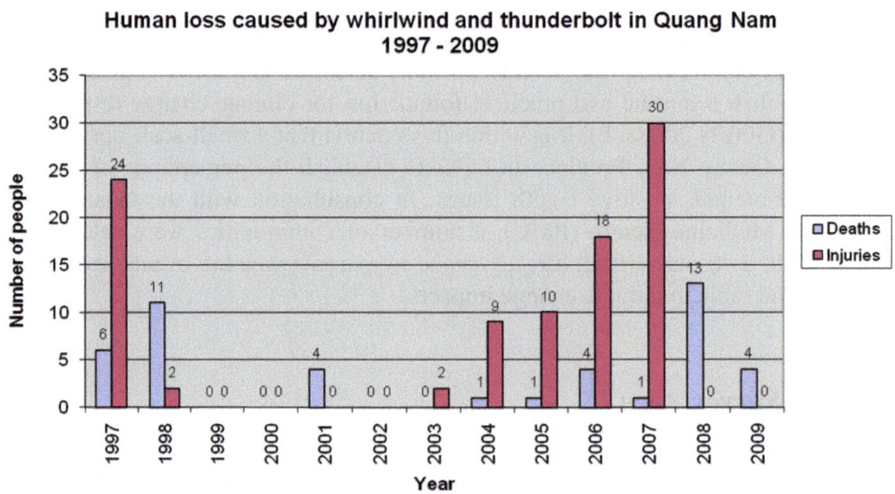

Fig. 5 Human loss caused by whirlwind and thunderbolt in Quang Nam 1997–2009 (Bình et al. 2011)

Minh commune has been selected as a study area out of the 21 communes emphasizing a coastal area with past history of extreme weather events (see map Fig. 7). Binh Minh covers an area of 1,180 ha and has a population of approx. 7,100.

The entire province of Quang Nam has witnessed a series of storms and floods in recent years. Particularly the years 1999, 2006, 2007 and 2009 saw considerable numbers of fatalities and injuries from weather events, as seen in Fig. 4.

Other extreme weather events have impacted local communities in Quang Nam Province. Figure 5 illustrates the impacts of whirlwind and thunderbolt in the

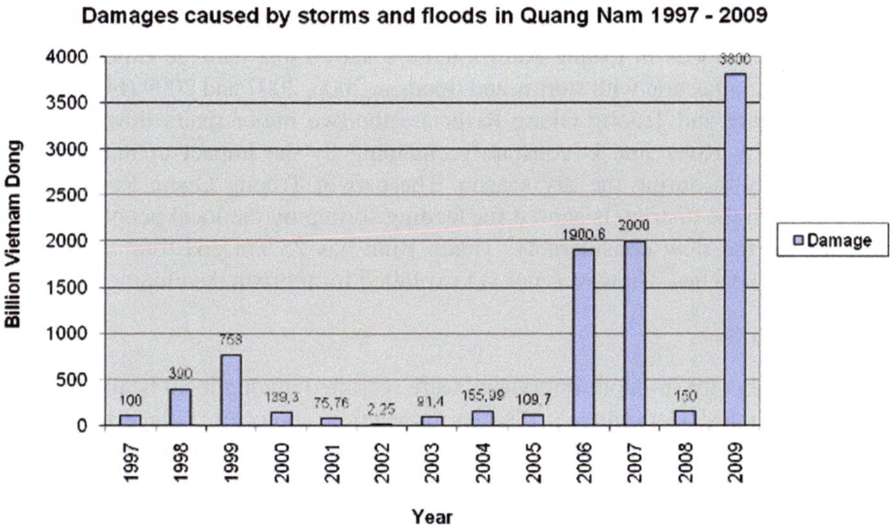

Fig. 6 Damages caused by storms and floods in Quang Nam 1997–2009 (Bình et al. 2011)

Fig. 11.7 Map of Quang Nam province indicating location of Binh Minh

same period of 1997–2009. Fatalities have occurred in nine out of 12 years, and increasing injuries were particularly seen from 2002 to 2007.

Similarly, tremendous losses have been sustained due to physical damages caused by storms and floods in Quang Nam. Figure 6 shows that damage experienced in particular in connection with storms and floods in 2006, 2007 and 2009 (Fig. 4).

Ly Ly River and Truong Giang River are the two major rivers flowing in the district. Ly Ly River line is constantly changing by the impact of major floods and often dry-in during the dry season. The part of Truong Giang River, which flows through the district, is shored for feeding shrimp by the local people, thereby diminishing the flow considerably. Thang Binh has 25 km coastline with many high quality beaches—however, not yet exploited for tourism development.

Key informants

Key informants for in-depth interviews were recruited among local health authorities, the Peoples' Committee's disaster prevention authority, village-based Red Cross, and a village representative from small-scale business and fisheries sectors. Interviews were conducted by means of guides with structured questions related to climate change awareness, health impacts and adaptation. At the time of the survey in July 2010, a series of focus group discussions were also performed.

Key observations

Awareness of climate change

Most of the respondents associated climate change with abnormal weather conditions and typically mention seawater level rise, storms, floods and increase in temperature. Typical replies by respondents reflected the opinion that climate change is a phenomena of weather that is not associated with the known natural regime: "*Climate change means the impact of the weather without a rule*", "*Irregular rainfall, it is not cyclical*", "*Generally it is unexpected rain or sun and more severe*". In the daily life of the fishing community the unpredictable storms mean that it becomes difficult for fishermen to properly schedule their work. A female respondent noted that "*Before the beach was far away when we went to collect snails, but now the shore is moving very close*". Another interviewee shared this view: "*Sea water level is now rising and it has caused flooding and broken dykes. The dyke was built in 1965—now water is close to the dyke*". Some respondents claimed they did not know about climate change, but were nevertheless able to list a number of associated phenomena such as more severe storms and higher temperatures. Some respondents clearly mentioned linkages to environmental and habitat destruction, to deforestation, and to global warming causing ice melting and a rise in the sea level. A rise in temperature and a decline in precipitation were mentioned by several respondents, resulting in more drought periods: "*It dries the body, it is formidable impact that people cannot bear. It is hotter than usual, making it more difficult to breathe and so on*".

Impact on health

Generally, respondents had observed considerable changes in health patterns in recent years. However, these were not immediately linked to climate or weather factors, but rather linked to changes in environmental determinants of health such as food, water and air quality. Of much concern was the apparent increase in instances of cancer: "*Many people have started to look at the reasons for more cancers in recent years in the village. From 1984 to 1990 it was not much. It has occurred a lot more since 1990... I saw many cases of cancer this year, the main cause is well-water quality. Some people have been affected by alum*". Another respondent stated: "*Compared with the earlier period, there has been a big change and now there is too much cancer –there are nearly 100 cases*".

Although it cannot be concluded that the cancer cases mentioned by respondents are climate change related, it may indirectly be a possibility. According to Minh et al. (2008), in agriculture there has been widespread use of organochlorine insecticides, in particular DDTs (so-called 'persistent organic pollutants' or POPs) in Vietnam and high concentrations have been found in samples. A review by UNEP/AMAP (2011) confirms the linkage between DDT and a series of population health effects, including various cancer types such as breast, prostate and testicular cancer. The review furthermore highlights relationships between climate change, weather variability and POP levels. Extreme weather events such as flooding may release and re-distribute high amounts of pollutants to the environment with subsequent accumulation in food chains. Here children represent a particular vulnerable group due to their developing physiology and immune system as well as due to their comparable higher exposure to POPs relative to weight during eating, drinking and breathing. However, a potentially confounding factor to climate change is the evidence of an association between herbicides, including the widely used Agent Orange during the Vietnam War, and various types of cancers such as sarcomas, lymphomas, Hodgkin disease and leukemia (American Cancer Society 2012).

Of similar concern is the situation for diarrheal diseases. Several informants mentioned cholera as among the top five health concerns possibly related to climate and/or environmental change: "*Some diseases are very unusual and have not been seen before, for example a strange epidemic disease causing diarrhea in the whole community, and it is also strange to the Ministry of Health*"; "*There are many reports about strange cases of illness, which cannot be diagnosed. I think it is more or less due to climate change—a direct impact to health*"; and "*Just recently diarrhea cases have developed all over the district*". Particularly in relation to problems of overflowing latrines during flooding, a possible risk factor in relation to cholera, studies need to be carried out. Other water-related diseases mentioned included skin and eye diseases, usually having lack of water for personal hygiene as an important risk factor.

Dengue fever was referred to by several key informants as a rising health issue. An interviewee mentioned that, "*The second most important climate change related disease (after cholera) is dengue fever—it spreads so fast, and if not treated in time it is very dangerous, and in crowded communities it is a special concern*". One recalled that, "*The dandy (dengue) fever was epidemic in the rainy season about 3 years ago*".

Fatalities caused by drowning and severe injuries were also of much concern to the interviewees. The consequences of such for households' economy and employment as well as the derived mental problems in the form of post-traumatic stress were mentioned in several interviews. For instance, "*The Chan Chu storm was at first predicted to go in a southwest direction, and the fishermen believed so. When it went to the north and stroke, 86 people were killed in Binh Minh commune. There is a household where both father and son died*". Others stated, "*As for storm no. 9 (Ketsana), I am now 60 years old and I never seen so huge waves—there are storms every year, but not with such big waves. Its severity caused huge accidents—boats were pushed up and broken*"; "*Death matters the most, it has consequences for the parentless children*". Overall the impacts of extreme weather events are strongly recognized as being of major importance for peoples' life, livelihoods and living conditions. As one informant stated, "*Climate change affects health—but it also affects the lives of 90 % of the fishermen here. Climate change makes life harder, because there are sea temperature changes that make it more difficult to catch fish*".

Other observations

Referring to the wider socio-economic household survey undertaken in 2010 in selected coastal and mountainous districts in Quang Nam province (see "Impediments to Climate-Induced Disaster Management: Evidence from Quang Nam, Central Vietnam", "Livelihood Stresses Under the Constraints of Climate Change Vulnerability in Quang Nam" and "Rural households: Socio-Economic Characteristics, Community Organizing and Adaptation Abilities"), the above statements are affirmed by the health status of the respondents. Table 7 shows a

Table 7 Health today compared to 10 years back by district

			Health today compared to 10 years back			Total
			Worse	Same as before	Better	
District	Bac Tra My		18	3	5	26
	Dai Loc	Count	12	2	3	17
	Hoi An	Count	12	10	3	25
	Nui Thanh	Count	23	8	7	38
	Que Son	Count	28	6	3	37
Total		Count	93	29	21	143

Source Household survey

Table 8 Prioritization of factors with perceived impact on livelihoods in study communities

		Importance of bad weather/ natural disasters for livelihood	Importance of pests for livelihood	Importance of diseases for liveli-hood	Importance of agr. prices for livelihood	Importance of agr. input prices for livelihood	Importance of poor health
N	Valid	135	47	67	96	64	52
	Missing	31	119	99	70	102	114
Mean		2,11	3,26	3,34	2,92	3,17	3,13
Std. Deviation		1,201	1,310	1,200	1,262	1,292	1,314

Source Household survey

Table 9 Pay for health services?

			Pay for health services?				
			No answer	Na.	No	Yes	Total
District	Bac Tra My	Count	7	11	7	7	32
	Dai Loc	Count	9	7	1	7	24
	Hoi An	Count	0	6	1	18	25
	Nui Thanh	Count	10	5	14	19	48
	Que Son	Count	0	8	15	14	37
Total			26	37	38	65	166

Source Household survey

worsening of health conditions reported by a total of 93 households, an improvement reported by 21 households, and no change reported by 29 households.

Questions about stress factors considered to impact livelihoods were also addressed to households. Table 8 shows that in a ranking of the five most important factors, weather conditions and natural disasters were considered of the highest importance (mean 2.11) and poor health was ranked in third place (mean 3.13).

The results, although not fully representative, may indicate a higher confidence in the availability of health services than in the existing systems for weather forecasts, disaster warnings and disaster risk management. The section below will elaborate on some coping mechanisms as seen in the coastal district of Thang Binh.

Health equality and coping with changes in health and environment

Universal accessibility to health services is crucial for safeguarding human health, particularly in the context of uncontrolled development and climate change. It represents a challenge when health systems and services are put under stress. The socio-economic household survey mentioned above indicated that a considerable number of households pay for health services (Table 9).

Although the survey had a considerable number of non-respondents on health issues, the results may reflect a general problem since an estimated 55.5 % of the sources of total health expenditure in Vietnam are covered by private out-of-the-pocket (OOP) payments (Tien et al. 2011). During the implementation of the *Doi Moi* policies, the health financing mechanisms were reformed and a transition was made from a tax-based system to a system with multiple sources of funding, including government revenues, social health insurance, and out-of-the-pocket (OOP) payments (Tien et al. 2011). The results indicate that health inequality is still an important issue, potentially to be increased by a climate change induced burden to the health system and services.

When interviewing local residents in Thang Binh about their coping mechanisms, there was a sense of great social cohesion and willingness to help each other when threatened by bad weather or climate change impacts: "*When flooding occurs, most neighbors help each other*"; "*I think local people are enthusiastic and eager to help—uploading sand bags on roofs to keep them from blowing away*".

A whole range of coping mechanisms were highlighted in times of increased extreme weather events and general environmental change. The building of new strong houses and reinforcement of old houses were mentioned as key to disaster prevention (see "Climate Change, Adaptation and the Environment in Central Vietnam" and "Rural households: Socio-Economic Characteristics, Community Organizing and Adaptation Abilities"). Underground shelters for use by several households were emphasized in some localities, and the need for proper sanitation facilities and safe water supply to maintain good hygiene in protection against e.g. diarrheal diseases, were mentioned in other places. Planting trees for protection and shade were also considered important, just as were the dredging and maintenance of ditches to allow for proper water management. Finally, a wish for an efficient early warning system was mentioned, including reliable weather forecasts, in order to ensure overall disaster preparedness.

Health system responses to climate change

The health clinics at communal level have a major responsibility for the annual planning of flood and storm prevention measures. They have established mobile emergency teams as well as systems for monitoring the local health situation. There are good linkages to higher-level health facilities and authorities, which provide the necessary guidance and assistance such as for water treatment and environmental sanitation. Some of the challenges reported relate to understaffing in the health sector and lack of specific drugs, materials and equipment.

Local communes have put considerable work into the establishment of Flood Control Committees. These have an important role in creating awareness about climate change and disaster related issues and they participate in collaborative actions with the Youth Association and the Marine Corps, in order to assist fishing communities, for example, with rescue operations and teaching swimming skills. Such activities are in line with the overall National Target Program in Response to Climate Change (2008) and the action plan developed by the Ministry of Health in response to climate change (GOVN 2010).The action plan has among its objectives the evaluation of climate change impacts on health and the broadening of awareness among health staff and communities on health protection and climate change adaptation.

6 Conclusion

'There is no lack of well thought out policies on climate change and disaster risk management in Vietnam—over 200 laws, policies and strategies exist! The overriding challenge being faced by Vietnam is implementing these policies effectively.' This is one of the key findings of a recent review of the existing policy framework on climate change and disaster risk reduction by AMDI and TPGC (2011). Even though this chapter has only presented very preliminary, small-scale studies not necessarily representative for the broader Vietnamese society, the main conclusion remains the same as above: for community-based climate change adaptation initiatives to be sound and sustainable, there is a need to implement existing policies, strategies and action plans effectively through joint, inter-sectoral action. The latter is a challenge in itself, because most people seem to associate risk management actions with the state sector rather than with the local community (see "Impediments to Climate-Induced Disaster Management: Evidence from Quang Nam, Central Vietnam"). This challenge may be overcome if action is first taken at community level that is capable of demonstrating positive and viable results. This may eventually benefit the most vulnerable population groups, who still show little awareness of the inter-linkages between health and development-induced environmental and social change. Ideally, new initiatives should be informed by more substantial research; the present material may serve to stimulate both more in-depth and broader studies to identify proper, sustainable measures to cope with climate and environmental change.

References

ADB (2009) The economics of climate change in south East Asia: a regional review. Asian Development Bank, Mandaluyong City

Ahn TTV (2009) Gender and climate change in Vietnam: A desk review. United Nations—Vietnam

American Cancer Society (2012) Agent Orange and cancer. www.cancer.org/accessed Nov 2012

AMDI, TPGC (2011) Climate change and disaster management policy in Vietnam. Asian Management and Development Institute Vietnam and The Pressure Group Consultancy, UK

Baumann LC, Blobner D, Binh DT, Lan PT (2006) A training program for diabetes care in Vietnam. Diab Educ 32:189. doi:10.1177/0145721706286569

Beran D et al (2009) Report on the rapid assessment protocol for insulin access in Vietnam. International Insulin Foundation, London

Berry HL, Bowen K, Kjellstrom T (2010) Climate change and mental health: a causal pathways framework. Int J Pub Health 55:123–132

Bình TTT, Vân DK, Furu P (2011) Nghiên Cứu Bước Đầu Về Các Tác Động Sức Khỏe Và CơChế Ứng Phó Với Sự Biến Đổi Khí Hậu Ở Một Số ĐịaPhương Có Nguy Cơ Cao Tại Miền Trung Việt Nam.Báo Cáo Chuyên Đề 10.DỰ ÁN P1-08 VIE

Confalonieri UB et al (2007) Human health. Climate change 2007: impacts, adaptation and vulnerability. In: Parry ML et al (eds) Contribution of working group II to the fourth assessment report of the intergovernmental panel on climate change, Cambridge University Press, Cambridge, pp 391–431

Few R, Tran PG (2010) Climatic hazards, health risk and response in Vietnam: case studies on social dimensions of vulnerability. Global Environmental Change. doi:10.1016/j.gloenvcha.2010.02.004

GOVN (2006) Comprehensive development design for the health system in Vietnam to 2010 and vision by 2020. Government of Vietnam, Hanoi, p 49
GOVN (2007) National strategy for natural disaster prevention, response and mitigation to 2020. Government of Vietnam, Hanoi
GOVN (2008a) National target program to respond to climate change. Government of Vietnam, Hanoi
GOVN (2008b) Decision No. 158/2008/QD-TTg dated 2nd Dec 2008 on approval of the National Target Program to respond to climate change. Government of Vietnam, Hanoi
GOVN (2010) Action plan. Response to the global climate change period 2010–2015. Accompanying decision No. 3557/QD-BYT dates 27 Sept 2010 of the Minister of Health. Ministry of Health, Government of Vietnam, Hanoi
GOVN (2011) National climate change strategy (issued with the Decision No. 2139/QD-TTg dated 05 Dec 2011 of the Prime Minister). Government of Vietnam, Hanoi
Ha DA, Chisholm D (2011) Cost-effectiveness analysis of interventions to prevent cardiovascular disease in Vietnam. Health Policy Planning 26:210–222
Hai VM (2004) A study of gender relations in disaster management in three disaster prone areas of Viet Nam. Vietnam Red Cross and International Federation of Red Cross and Red Crescent Societies
IPCC (2001) Climate change 2001: synthesis report. A contribution of working groups I, II, and III to the 3rd assessment report of the intergovernmental panel on climate change. In: Watson RT, The Core Writing Team (eds) Cambridge University Press, Cambridge
ISPONRE (2009) Vietnam assessment report on climate change. Institute of Strategy and Policy on Natural Resources and Environment, Hanoi
Jørgensen LB (2011) Impact of typhoons on persons with Disabilities: a field study in the Guang Nam Province of Central Vietnam. Master of Disaster Management Thesis. University of Copenhagen, Denmark and University of Lund, Sweden
Kien TM, Hanh TTT, Cuong HC, Shaw R (2010) Identifying linkages between rates and distributions of malaria, water-borne diseases and influenza with climate variability and climate change in Vietnam. Environ Disaster Risk Manag 5:417–449
Kovats RS, Menne B, Ebi KL (2003) Methods of assessing human health vulnerability and public health adaptation to climate change. WHO Regional Office for Europe, Health and Global Environmental Change Series (1), Copenhagen
Lan TTN, Anh LM, Trang KTQ, Le PTT (2011) The situation of drowning mortality in Vietnam 2005–2009. In: Proceedings of the world conference on drowning prevention, Abstract, Hanoi, p 39
Millennium Ecosystem Assessment (2005) Ecosystems and human well-being. Synthesis. Island Press, Washington
Minh HV, Huong DL, Giang KB, Byass P (2009) Economic aspects of chronic diseases in Vietnam. Global Health Action (2) DOI:10.3402/gha.v2i0.1965
Minh TB, Iwata H, Takahashi S, Viet PH, Tuyen BC, Tanabe S (2008) Persistent organic pollutants in Vietnam: environmental contamination and human exposure. Rev Environ Contam Toxicol 193:213–285
MOH (2010) Five year health sector development plan 2011–2015. Ministry of Health. Government of Vietnam, Hanoi
Narain JP, Garg R, Fric A (2011) Non-communicable diseases in the South-East Asia region: burden, strategies and opportunities. Natl Med J India 24(5):280–287
Ngoan LT, Lua NT, Hang LT (2007) Cancer mortality pattern in Vietnam. Asian Pac J Cancer Prev 8(4):535–538
Nhung NTT, Long TK, Linh BN, Vos T, Anh ND, Huong NT (2011) Vietnam burden of disease and injury study 2008. Medical Publishing House, Hanoi
NCADPPC (2012) Vietnam AIDS response progress report. Following up the 2011 political declaration on HIV/AIDS. National Committee for AIDS, Drugs and Prostitution Prevention and Control, Hanoi
O'Brien KL, Leichenko RM (2000) Double exposure: assessing the impacts of climate change within the context of economic globalization. Global Environmental Change 10:221–232

Tien TV, Phuong HT, Mathauer I, Phuong NTK (2011) A health financing review of Vietnam—with a focus on social health insurance. World Health Organization, Hanoi

UNEP/AMAP (2011) Climate change and POPs. Predicting the impacts. Report of the UNEP/AMAP expert group. Secretariat of the Stockholm Convention, Geneva

Vuong DA, Van Ginneken E, Morris J, Ha ST, Busse R (2011) Mental health in Vietnam: burden of disease and availability of services. Asian J Psychiatry 4:65–70

WHO (1997) Health and environment in sustainable development. Five years after the earth summit. World Health Organization, Geneva

WHO (2003) Climate change and human health - risks and responses. Summary. World Health Organization. ISBN 92 4 159081 5

WHO (2009) Vietnam country profile of environmental burden of disease. World Health Organization, Geneva

WHO (2010) Western pacific country health information profiles: 2010 revision. World Health Organization Regional Office for the Western Pacific, Vietnam, pp 490–506. ISBN 978-92-9061-491-3

WHO (2011a) Western Pacific Country Health Information Profiles: 2011 Revision. Viet Nam, pp 463-472, 490-534. World Health Organization Regional Office for the Western Pacific. ISBN 978-92-9061-540-8

WHO (2011b) World health statistics 2011. Part II. Global health indicators. World Health Organization, Geneva

WHO (2011c) NCD country profiles, 2011: Vietnam. World Health Organization, Geneva. http://www.who.int/nmh/countries/vnm_en.pdf/. Accessed July 2012

WHO (2012) Emerging disease surveillance and response: Avian flu. World Health Organization, Geneva http://www.wpro.who.int/emerging_diseases/AvianInfluenza/en./. Accessed July 2012

World Bank (2008) Global monitoring report. MDGs and the environment. Agenda for inclusive and sustainable development. World Bank, New York

Yusuf AA, Francisco H (2009) Climate change vulnerability mapping for South East Asia. Economy and Environment Program for Southeast Asia (EEPSEA) http://www.eepsea.org

[illegible]

Interacting Cultural and Environmental Change: The Co (Cua) Minority of Central Vietnam

Nhinh Do Thi and Ole Bruun

Abstract This chapter shows how dramatic changes in the highland forest environment have impacted on a mountain people whose existence since prehistoric times has been bound to the forest. It combines scarce textual sources on their culture and history with recent fieldwork in their lands. This is in order to explore the process by which climate and particularly environmental change have contributed to social stratification and generated new vulnerabilities. An ingrained antagonism between Co people and shifting lowland state formations, which have repeatedly submitted them to their rule, forms part of the region's history. Yet the post-war changes in the forest environment, combined with government integration efforts, have been the most serious challenge to their traditional knowledge and way of life and have ultimately threatened their cultural survival. At the same time, the retreat of the natural forest has reshuffled resource entitlements in the highlands and exposed the group to renewed competition for land and resources.

Keywords Environmental change • Forest • Co (cua) people • Food production • Cultural identity • Tensions between ethnic groups

N. D. Thi
GNI, Hanoi, Vietnam

O. Bruun (✉)
Department of Society and Globalisation, Roskilde University, Roskilde, Denmark
e-mail: obruun@ruc.dk

O. Bruun and T. Casse (eds.), *On the Frontiers of Climate and Environmental Change*, Environmental Science and Engineering, DOI: 10.1007/978-3-642-35804-3_12,

1 Introduction

The small group of people known as the Co (or Cua, Cor, Col among others[1]) inhabits the previously densely forested highland area across the Quang Nam and Quang Ngai border in South Central Vietnam. The center point of their territory is usually considered to be Mount Chua in Quang Nam. They belong to the Mon-Khmer people, who inhabit Cambodia and a long mountain corridor up through present Vietnam and Laos. During the Khmer empire (802–1431), the Co lands formed part of the border area with the lowland Champa Kingdom. Their language belongs to the Central Bahnaric subgroup, with relatively close relations to that of neighbouring peoples like the Hre, Xo-dang and Ba-na, but is itself divided into several dialects. Their numbers were estimated at 20,000 in the 1960s (Phillips 1962; Smith 1965; Hickey 1982a, p. 482), close to the official count of 22,649 in 1989 which in terms of size places them in the mid-range of the 53 recognized minority groups in Vietnam. In 2009 a total of 33,817 people were registered as belonging to the Co minority (VCDS 2009), now spread across 25 provinces; of these 5,361 lived in Quang Nam and 28,110 in Quang Ngai, the two provinces of their native area. After the Doi Moi reforms, considerable population growth has occurred among the Co as has been the case among Vietnam's other ethnic minorities (ADB 2002, p. 8).

The Co are among the least known of the Vietnamese mountain people. Few written sources about them are available and they are not themselves known to have produced written material on their history and way of life.[2] A few general works on the minority peoples of Vietnam provide popular introductions to Co culture (e.g. EVC 2012; Dang et al. 2000). The bulk of the available material originates from before the American war, these are general French era 'Montagnard' studies (see Salemink 2003), or from the American era. Ethnographic material was collected by Jackie Maier and Eva Burton of the Summer Institute of Linguistics as well as by the anthropologist Robert Mole (1970), and many references to the Co are found in the comprehensive work on the highland minorities by American anthropologist Gerald Hickey (Hickey 2003b)[3]. Other accounts were assembled for US Army Special Forces training purposes (Schrock et al 1966; USASWS

[1] The Co are also known as Cor, Col, Cùa, Trầu and Moi in Vietnamese sources and as Cua, Khua, Kor and Traw in Western sources. Early research by SIL suggests that among themselves they refer to those at higher elevations as Kor, Koh, Dot, and Yot, and those in the valleys as Traw and Dong (Hickey 1982a, p. 26). Thus, the Co hardly existed as a clearly discernible 'tribe' in the pre-modern era [tribes may in may in fact be a common fiction (e.g. Leach 1954, p. 290)], but rather as a collection of cultural groups and dialects with close association.

[2] In the 1960s, efforts were made by Jackie Maier and Eva Burton of the Summer Institute of Linguistics (a Christian missionary organization) to create a written Co language based on Latin script, but is has not been sustained after 1975.

[3] Apart from historical material, Hickey based his account on first-hand knowledge of two SIL researchers working with the Co language (see note 2) and interviews with Co people in Tra Bong in 1966.

1965). Scattered references to the Co are also found in missionary literature from the mid-twentieth century (e.g. Smith 1965). In recent times a small book on the Co was assembled by a Vietnamese anthropologist (Chu 2004), based on Vietnamese documents and personally collected material, and another book in Vietnamese has a collection of traditional Co stories (Vu 1992).

In addition to these general sources, the present chapter builds on a long range of interviews with Co people and field observations in their villages during field trips over the period 2009–2011. The comprehensive household survey in Quang Nam, which several chapters of this book draw on, included seventeen Co households in two villages. This provides some relevant socio-economic data and a comparison with the majority population.

2 Environment and Overview

The Co people have practiced slash-and-burn, rotating cultivation on mountainous forestland throughout historical time. The territory they inhabit is an extension of the Annamite Range (Dãy Trường Sơn), protruding towards the coast and forming a natural divide between present Quang Nam and Quang Ngai provinces. The ground rises and falls between altitudes of a few hundred meters and up to 1,383 meters at the highest peak (Mount Chua), with several ridges rising above 1,000 m. There is little plain ground in this rough mountain terrain, which mostly rises steeply and is cut by many narrow and steep river valleys with swift-flowing streams. The central ridges of the Co territory are the sources of many water courses, flowing off in all directions and feeding the present Phu Ninh reservoir in the north, the Thu Bon river in the northwest and Tra Bong River to the south. These surrounding river valleys and lower hills, where level ground is found, also, historically speaking, form part of the Co lands, which comprise as much as 1,200 km^2 (Mole 1970, p. 177).

The history of the Co, what is known of it, reflects their central position in the perpetual struggle between Cambodia, Champa and Annam since prehistoric times, interrupted by the intrusion of Chinese and Mongol forces. The rich resources in their mountain lands, including hardwood, ivory, cinnamon and tea, incited political domination by shifting state formations. At least in the eleventh century, and perhaps much earlier, they were subjected to the Champa Kingdom, which controlled Central and South Vietnam and had several important centers, including one of its capitals in present day Quang Nam (several archaeological sites remain today). As a seafaring and trading people, the Cham extracted cinnamon and other commercial goods from the Co territory. The pressure, initially, eased when under the Emperor Le Thanh Tong (1460–1497) of the Le Dynasty ethnic Vietnamese expanded south and defeated the Champa in Central Vietnam in 1471 (Le et al. 1993), because the Viet only peripherally engaged with the mountain people. Instead, guard posts and military colonies were established where Co lands bordered on the plains. However, soon after, highland people there were brought under the control of Bui Ta Han (1496–1568) who became governor of Quang Nam[4] in 1540. Precious forest products, including cinnamon, elephants, ivory, rhinoceros tusks, hardwood, wax, rattan and betel, were objects for long-distance trade (Hickey 2003a, p. 2) in which the Chinese and possibly Japanese and Portuguese also soon engaged. In the subsequent period, trade was primarily performed by merchants with special privileges for a given minority area on condition that they simultaneously extracted tributes from it. The so-called *cac-lai* traders tended to consist of social outcasts who became known for their reckless behavior and excessive demands, which aroused violent reactions from the highlanders (Condominas 1951, p. 79). A common Vietnamese denomination of all highland peoples as Mọi, 'savages', is evident from this period onwards, which has left enduring antagonisms. Under the Nguyen rulers from 1802 increasing Vietnamese immigration into the lower highlands of Quang Nam resulted in the establishment of trade settlements at Tra My and Phuoc Son, both of which became 'flashpoints of conflict between the Cua [Co] and dishonest lowland merchants' (Hickey 2003a, p. 3). After 1820, a new ruler made cinnamon trade a royal monopoly under strict administrative control, which affected a sharp decline in Vietnamese settlements and improved relations between Co and Vietnamese, yet without ending trouble.

After 1862 the Co were again subjected to attempts at pacification and trade monopolization. The *Son Phong* scheme of this period implied the construction of a line of fortifications from Tra My to Chi Doc to control the highland people as well as their resources, not least the valuable cinnamon. Despite the existence of markets, most direct trade took place outside, performed by *Son Phong* agents with sublet rights to both tax collection and trade. Serious conflicts and incidences of plunder prevailed until the French dismantled the *Son Phong* at the end of the nineteenth century, instead making French officials responsible for the trade through to 1954. French colonial administration was based on regional divisions (Hickey 1970, p. 10), including the segregation of lowland and highland people. In 1950, the French established an autonomous Pays Montagnard du Sud (PMS) under the authority of the emperor Bao Dai whom the French had installed as an alternative to Ho Chi Min in the North. Political events of the time, including the Geneva Accord of 1954 effecting vast population movements of both minorities from the North and lowland Vietnamese into the central highlands, and a return to Vietnamese political dominance,[5] inspired a further push for independence among the highland people. As early as 1956 a new land reform program aimed at the transformation of the agricultural practises and lifestyle of highland people towards sedentarization (Hickey 1982b, pp. 28–29), and land ownership decrees from 1958/1959 denied them title to their lands. Movements for autonomy occurred from 1957. In 1964 the FULRO

[4] Quang Nam is the former name of a region in the Central Vietnam covering Da Nang, Quang Nam, Quang Ngai, Binh Dinh and a part of Phu Yen today.

[5] The assimilation policy of president Ngo Dinh Diem's (1955–1963) had a clear element of Vietnamese ethnic chauvinism and a wish to eliminate French influence in the highlands (Hickey 1982b, p. 9–11).

(United Front for the Liberation of Oppressed Races) was established, aiming at full independence for the highlands.[6] Reactions from Vietnamese government and military officials were harsh. While promising to correct all wrongs of the former regime (under Diem) and improve the life of the highlanders, government policy emphasized the 'Great Family of our Nation' and vowed to 'take any hard measures… to eliminate all conspiracies and activities prejudicial to national sovereignty and authority' (Vinh 1965, pp. 29, 38).[7]

The memories of the eldest Co people reach back to the 1930s, but tend to be more consistent from the 1950s where they complement written sources. Memories of trade relations, vital to the Co people, occupy a prominent space. Elderly Co people generally agree on the exploitative practises of lowland traders. They say that the illiterate Co did not understand the denominations of Piaster notes and were easily cheated. Or they provide many examples of the poor deals they were given when they, apparently still unable to trade directly in lowland centers, were met half-way up the mountain by Vietnamese brokers. Mr. Lê Văn Phúc, a former sub-village leader (born around 1933 though the exact date is unknown) remembers how they exchanged several kg of tea for a single kg of rice (it starts a debate and other villagers say up to 7 kg of tea was exchanged for 1 kg of rice). They remember how, with the establishment of the Republic of Vietnam in 1955, the Vietnamese again had taken control of the highland trade while the political turmoil in this period opened up for ruthless practises. According to villagers in Tam Tra, the Co people were not permitted to carry firearms, did not have unrestricted access to lowland markets, and were generally shunned by the Vietnamese in this period. An infamous example concerned the General Van Toan (nicknamed the Cinnamon General) of the Second Division stationed in Co lands who was notorious for helping himself to cinnamon and selling it on the black market (Hickey 2003a, p. 9).

With war looming on the horizon, however, the Co people were increasingly engaged by all warring parties due to the strategic position of their lands. In the late 1950s and early 1960s loyalist to North Vietnam began infiltrating the higher Co lands, making use of Co people that had moved with the Viet Minh to Hanoi after the Geneva Accord (the Quang Ngai/Quang Nam perimeter was a provisional assembly areas for Vietnam People's Army and these provinces were already Vietminh strongholds) (Hickey 1970, p. 40; Fall 1963. p. 193) and been had trained as cadres at the Southern Ethnic Minorities School outside Hanoi. Some sources mention that these had gained considerable influence at the expense of government control (USASWS 1965; Le 2004). Certainly Quang Nam was a

---

[6] The FULRO movement, led by Bham Enoul, had land rights as a central concern. It formed a provisional government in exile, attempted to form its own military force, and on several occasions made armed insurrections against the South Vietnamese army. It further made pledges for autonomy to the United Nations and appeals for support to the American embassy in Saigon.

[7] This source at the same time reflects conventional chauvinism in statements saying that the highlanders possess no common ideology or community spirit which could unite them into larger groups; that although their existence has lasted for a very long time, nothing remains as a proof of a cultural system; and that they live from a very primitive agriculture and mostly from resources found in the forest (Vinh 1965, p. 11).

major entry point for NFL/Vietcong infiltration with way stations a day's March apart (Raskin and Fall 1965, p. 146), and according to locals they greatly affected Co lands.

An increasing American military engagement in the highlands had been preceded by American missionary work among ethnic minority areas over several decades. Seeing the foreign power as a possible support for their cause or at least a 'buffer' to the Vietnamese (Fall 1966, p. 191), a great many highland people had built loyalty to it.[8] In 1962, the US Special Forces opened a Camp at Tra Bong and started recruiting young highlanders for militia, including many Co people (special warfare, porters and scouts), and another camp was established at Tra My in 1964.[9] When the Special Forces were integrated into the Vietnamese Army corps in 1964, however, desertions and rebellions followed (Fall 1966, p. 195).[10]

The animosity between South Vietnamese and highland people continued to be a major issue during the war; not until 1966 were South Vietnamese soldiers and staff instructed to refer to the highland people as 'compatriots of the highland' (Dong Bao Throng) as opposed to *Moi*. This was done in an effort to win them over (Kelly 1973, p. 20). Co people consistently speak of not being recognized as a minority, but only being designated *Moi*, even as *Moi có đuôi* (wild animals with tails) by lowlanders throughout this period. Another point of contention was personal names. The Co had their native names, though traditionally not much used (certain taboos applied) and parents mostly used standard terms for their sons and daughters. But a very complex kinship terminology emphasized kinship relations and bore a strong taboo against common-ancestor marriage in the male line six generations back. Vietnamese names were commonly taken in adulthood. Vietnamese authorities may, as early as the sixteenth century have mandated, a standard surname of Đinh (uncommon in Vietnamese and allegedly indicating inferiority) for all Co people, presumably for tax collection purposes. When possible in the 1960s, many Co people chose common Vietnamese names instead. For instance, Lê Văn Phúc, mentioned above, was previously named Lê Văn Đinh, which again was put on top of a Co name.[11]

The period 1955–1963, under President Ngo Dinh Diem's nationalist and autocratic rule, saw renewed efforts to assimilate the highland people into Vietnamese culture. As an unofficial 'civilizing process' it implied imposing Vietnamese

[8] 40,000 highland people later fought alongside American forces of which an unknown number were Co.

[9] Both camps were relatively short-lived. Tra Bong was turned over to South Vietnamese forces in 1963, but in 1965 a new US camp was established there (Hickey 1982b, p. 9). Tra My was opened in 1964.

[10] One of the most notorious was the Ban Me Thuot rebellion, in which highlanders massacred Vietnamese officers, raised a highland minority flag and used radio transmitters to clamor for independence.

[11] Le Van Phuc is actually his son's name, used to avoid taboo. It is uncertain if this is actually a Co taboo or if it originates from Kinh people through Vietnamese soldiers. Some Kinh people in the North used to call people by their first child's name, while real names were only used when young or after death.

social institutions and cultural characteristics on them. At the same time the Land Development Program infringed on their land and traditional agriculture by resettling lowland Vietnamese in the highlands, while a Highland Resettlement Plan forced many off their ancestral lands and into 'reservations' (Hickey 1970, pp. 24–25). However, these initiatives ceased in Quang Nam during 1964 due to corruption, internal strife and increasing Vietcong control, and also a foreign-funded Strategic Hamlet Program for rural development came to a halt (Tanham 1966).

The highland peoples, which presumably like most of the political groupings in the politically fragmented South Vietnam were inherently neither pro-NFL[12] nor committed to the government (Hickey 1970, p. 6), were increasingly drawn into national and international politics. While the Bajaka and FULRO movements were fueled by nationalist politics and stretched into Cambodia and Laos, the NFL/ Vietcong saw fertile ground for infiltration in the highlands. Some reports suggest that the Vietcong promised the minorities an autonomous status once they were liberated (Hickey 1982b, p. 113), but in reality the North Vietnamese ethnic policy shifted to integration in the late 1960s (Masako 1999). New democratic institutions and many attempts at national accommodation in the post-Diem period, including official recognition of many FULRO demands, did not halt the militarization of the highlands.[13]

Increasing infiltration, the establishment of several army bases on Co lands, and the scaling up of the American-Vietnamese war was devastating for the small Co communities, both physically and mentally. Divided loyalties between the warring parties drew wedges into Co culture and in many cases covered confusion over the many promises made and distrust in the true motives of outsiders. Mostly concerned about protecting their way of life and securing their daily needs many Co people fought in the war (few of whom survived) or worked for one or the other party. As the war intensified, life became a matter of survival. For instance, Nguyen Thi Hong Cuong, born in 1955 and orphaned already at the age of 6, was recruited as a porter for the Vietcong to support their advance in 1971. She dug up bags of rice buried in underground posts and carried them across the mountains into the next province in order to deliver them to soldiers in the forest. Lê Văn Phúc, mentioned above, fought with the Vietcong and survived, later to become a village leader in the resettled village.

The Co traditional longhouses, forming the basic village communities, were apparently all shattered or burned to the ground during the war. Some were hit by bombs or caught fire from nearby bombings by American and South Vietnamese planes, and others were destroyed by the Vietcong as reprisals for not cooperating. As a survival strategy, many Co split up the longhouse village

[12] The NFL, also called Vietcong (Vietnamese Communists), saw its formal appearance in 1960 and worked for unification with Communist North Vietnam.

[13] In May 1961, American President Kennedy sent 400 U.S. Army Special Forces (Green Beret) troops into South Vietnam's Central Highlands to train Montagnard tribesmen in counterinsurgency tactics, and in 1965 Hanoi began to deploy into the South increasing units of the regular soldiers from the People's Army of Vietnam.

communities into smaller sections, each building a house for a smaller number of families, typically 2–5, and hiding them under the canopies of old forest to escape bombings. During the course of the war these were also devastated and a great many Co people speak of living merely in small bamboo huts, caves and crevices under large rocks under the most terrible conditions. As the war intensified, they had to move with increasing frequency; some report never staying in one place more than 10 days. Normal agricultural production broke down in the later years of the war and people fed themselves on wild plants, roots and whatever they could collect in the forest. By the end of 1964 the Vietcong had established effective control of the highlands in Quang Nam and reached the coast (Tanham 1966). Like other highlanders the Co could take to the bush to secure their liberty, but suffer infiltration and bombardments there, or they could seek protection around army camps, mostly to become beggars and hirelings (Condominas 1977, p. xv).

Lowland Co communities practicing paddy farming had in fact grown since the early twentieth century, when the French colonial administration had assembled Co longhouses in the lowlands into new administrative 'villages'. In the colonial administration a canton chief (*san*) had been appointed for a cluster of villages (Hickey 2003a, p. 17), and these 'chiefs' were still exercising authority according to tribal laws and customs in the 1960s. An interview with a Co man from this period reveals that his family had in fact been settled in the Tra Bong valley for seven generations (Hickey 2003a, p. 28). Several sources indicate that some Co people had settled on lower ground, either near the trading town of Tra Bong in Quang Ngai (e.g. Smith 1965, p. 43; Hickey 2003a, pp. 6–8), near the market town of Tra My in Quang Nam, or along the road from Highway 1 to Tra My (Mole 1970, p. 177). One Co man remembers that he already around 1960 moved into a lowland village consisting of three longhouses. Yet contemporary sources indicate that the majority of the Co people were slash-and-burn agriculturalists in the highlands; many of the Co people previously inhabiting market towns may to a higher extent have been assimilated into Vietnamese society.

3 Forest Life, Food Production and Trade

Highland forest products were cherished by lowland states and were invaluable for their long-distance trade. For the Co themselves, trade was essential for upholding the highland production system, but always involved the danger of simple cheating, tributary relationships and political domination. Highland peoples are often engulfed in majority narratives of original primitivity as a function of isolation, but history tends to show otherwise. In a grander perspective, the highlands were inhabited later than the lowlands, and the central highland massif of Southeast Asia was conceivably a 'shatter zone' as a function of state formation and collapse, inhabited by fleeing subjects as much as those group that deliberately avoided the slavery, conscription, taxes, corvée labor, epidemics and warfare of state-based civilizations (Scott 2009, p. 7).

Trade and shifting relations with lowland states were part and parcel of Co history, but frequently subjected to terms set by military force. Evidently, a high degree of isolation is possible and the semi-nomadic lifestyle may itself serve as defense strategy against political domination (Bruun 2006). Both missionary reports from the 1960s and oral accounts show striking differences between, for instance, secluded and suspicious Katu/Co-tu villages living behind palisades in remote mountains and using traps and poisoned spikes on their mountain paths to fend off visitors (Smith 1965), and more open lowland Co communities in organized trade relations with Vietnamese. Within a fairly short span of time or territory, we see highly diverse strategies depending on physical location, ethnic strategies and the nature of political power in the lowlands.[14] Likewise, internal specialization among the highland people induced some groups to act as intermediaries between highland and lowland markets.

The Co people conventionally cultivate the mid-level slopes at altitudes of a few hundred to several hundred meters creating two distinct belts of rainforest in their environment. In higher regions and on steeper slopes there is primary rain forest, with tress standing as tall as 40 m and when intact forming a continuous canopy. Below stands a middle level of smaller trees and on the ground is a third layer of seedlings and plants adapted to the dark rainforest floor. Epiphytes are very common at all levels of the forest, with an abundance of lianas and other climbing plants. Along watercourses, where sunlight penetrates, bamboo and rattan are luxuriant. The lower areas are mostly covered with secondary rain forest, which has developed after forest clearing, cultivation and abandonment of the land for a given number of years. Here, the trees are smaller but together with shrubs form an impenetrable growth. In some locations, the highest mountain slopes cultivated by the Co only stand with high grasses when fallow. A network of steep paths connected Co villages with their fields and with each other.

The combination of primary and secondary forests with open fields for shifting cultivation has supported a tremendous biodiversity. The old forest itself is home to hundreds of tree species, an untold number of plants and a wealth of insects, birds, reptiles and smaller mammals. Of the larger mammals, elephant, gaur, tiger, bear, leopard, deer, boar, porcupine, jackal, civet, otter, langur, macaque, gibbon and rhesus monkey have always inhabited the mountains and interacted with humans.

The shifting cultivation of the Co people yields a range of crops, primarily for consumption but also for barter trade. The secondary forest is cleared after laying fallow for a specified number of years depending on location, usually between two and five years. The larger trees and brushes are felled by the men and left to dry for a month in the dry season. Then the fields are burned in carefully managed fires, where the men prevent them from spreading into the forest. Highland dry

[14] Evidently some highland groups in the area relied on stone tools and simple, adapted technologies until recently (Danang Museum), while other groups had access to metal tools and weapons.

rice is a cherished stable crop planted on newly cleared land and may be planted for several years successively depending on soil fertility, often in combination with maize, bananas or other secondary crops like yams and manioc. After being sown in seedbeds and planted extensively on hill slopes by the women in spring, the rice is left to grow with only minor weeding and care. According to interviews, it may produce anything between 150 and 1,500 kg/ha depending on soil fertility and weather conditions, but presumably averaging 800 kg/ha. Compared to paddy rice, the area output of dry rice is low, but this is compensated for by a much higher labour output due to natural fertility of the soil (Kemf and Vo 1999, p. 166). However, because of the natural dependency on weather variability dry rice offers lower food security periodically making people more dependent on collection in the forest as well as making paddy rice from lowlands a common ingoing trade object. Still both taste and cultural identity have made the Co prefer highland rice as their stable crop to this day.

A long range of garden crops resembled those of lowland areas, including tomatoes, carrots, cabbage, squash, onion, chili pepper and beans. Apart from bananas, a range of fruits were grown around the longhouse or in orchards, including orange, grapefruit, guava, jackfruit, papaya, mango and coconut. Special crops like tobacco, cotton and betel were also grown, and the traditional exchange crops of cinnamon and tea were grown on the mountains.

Presumably encouraged by the relative security of the French era, hybrid cultural forms developed with the settlement of Co villages in lowland areas, both around Tra Bong and Tra My. They adopted paddy cultivation, often with a surplus for trade, and engaged in steady cinnamon, tea, rice and handicraft trade with the Vietnamese. These villages had grown to considerable size before the war and benefited from several secondary roads established in the French era, later to become impassable in the turmoil of war (Mole 1970, p. 179).

If political conditions were favorable, a higher degree of wealth and comfort may be had from permanent exchange relations with sedentary agricultural and trading communities. As shifting cultivators, the Co depends on traded goods to a great extent. These are above all metal objects and textiles, without which the Co would revert to a primitive state, but also the vitally important salt from coastal areas. Metal objects include tools, blades, containers and musical instruments (mainly gongs), but also precious metals for jewelry and storage of wealth. Some Co villages did weave textiles for the traditional dress: the women wore elaborate bodices and skirts, while the men might only use simple loin clothes for daily use (both men and women wore jewelry such as beads, bracelets and metal rings about their neck). Vietnamese cloth has long been more common trade object and through the early-to-mid-twentieth century Western and Vietnamese clothes became more common, especially for the men. Some pottery, particularly big jars for rice wine, was also traded for daily as well as ceremonial use and may still be preserved today.

Cinnamon (the bark of the cinnamon tree), exclusively produced by the upland Co, has historically been a particularly valuable exchange crop and a basis for long-distance trade in the hands of merchants, but also an object of government control and monopolization. Highland green tea, grown and picked extensively in the high mountains, was another stable object of exchange. Many elderly Co people indicate that both cinnamon and tea remained vital products exchanged not least for rice in the mid-twentieth century. Other pre-modern trade items as mentioned above declined in importance. Hardwood logging in fringe areas has taken place throughout history, such as by means of forced labour, but does not appear to have been a source of income for the Co; that is to say at least with the last century it was primarily performed by Vietnamese loggers. In the early-to-mid-twentieth century, when shifting governments established representatives in minority villages and encouraged schooling of children, at the same time they contributed to a slow process of Co population concentrations in market towns. In these, additional marketable products like weaved mats and baskets, crossbows and other handicrafts were added to the traditional Co products.

4 Political and Community Organization

Like so many other highland people, the Co previously lived in longhouses (*sech*), standing on poles above the sloping ground. The longhouse community, consisting of individual households, was internally egalitarian and functioned as an independent social, political and economic entity. Until the French era, no joint political organization linked the scattered Co longhouse communities. They were only connected by intermarriages, though usually cementing long-term alliances (Mole 1970, p. 186). Patrilineal descent and patrilocality formed the overall principles of organization, in effect making each longhouse a closely knit common-descent group. For this reason a strict taboo on common-ancestor marriage six generations back in the male line was upheld; still today people will avoid common ancestors in several generations back.[15] Longhouse communities did not even appear to have a name or designation, such as in a tighter political organization. Instead, longhouses might simply be referred to as the 'roof' of this or that chief: Pơo Lây ông (the Roof of Mr.) *Xúng*, *Pan*, *Lướuõi* etc. The position of chief was hereditary, but conditioned on the chief's ability to work hard and provide a good example for his longhouse. The chief (*dayalan*) was traditionally the only source of formal authority in the longhouse, but was invested with duties of arbitration, punishment, overseeing rituals and making decisions on when to move the longhouse.

Longhouses were built by columns and crossbeams of logs, bamboo and rattan floors and partitions, and a roofing of grass. They could contain anywhere from a dozen families up to a hundred (20–40 appears most typical), and extend as far as necessary. Some Co people mention houses stretching 100 m. The internal division of the longhouse was given by custom and consisted of a main room (*gu'l*) extended at one side through the entire length of the house and provided space for

[15] A description of Co kinship terms and structure is provided by Hickey (Ethnographic notes).

common work tasks such as weaving mats and baskets, making tools and weapons, husking rice and so forth. Jars, gongs, and implements stood along the walls, baskets and bundles of food and forest products were stored on rafters and tobacco and various food items hung from the ceiling. The common room also served ceremonial and religious purposes in marriages and rituals as well as sleeping space for the young unmarried men. The other side of the longhouse consisted of compartments for married couples, usually further divided to make rooms for single women or widowed elders. Each compartment had its own open hearth for preparing meals, cooking pots and kitchen gear was kept on one side. Blankets, clothes and private items were stored on the other side and crossbows (the traditional weapon, often with poison arrows) and knives hung on the wall.

Animism formed the common religious orientation as well as disease etiology of the Co people. Their world was inhabited by innumerable spirits, on which human life and wellbeing depended, and their anger and retribution were continually feared. Rocks and trees around the longhouse might be spirited and set with powerful taboos, and a sacrificial stake might be placed before the longhouse. Three ritual celebrations during the year implored the spirits to bring good health and abundant crops (Hickey 2003a, p. 25) by means of ritual sacrifices of chicken, pigs and buffalo respectively. The buffalo offering is the most important annual ritual, lasting 8 days. Previously centered on the longhouse community, it later developed into a larger communal event, gathering a number of longhouses. At these rituals gongs are played and women decorated with beads and flowers beat gongs and perform dances, each using her personal, creative style, while singing the old Co songs. At the buffalo offering, hot blood from the heart, kidney and stomach parts are offered to the spirits, and a range of rituals include the chief putting a dash of a bloody mixture on the head of each individual, calling on their spirits to return. Certain taboos restricted the use of personal names; children were not allowed to call their parents by name and children were commonly just called *Ôn* (boy), *Lung* (girl) or *Bẻem* (baby). Once reaching adulthood, age was not counted.

Domestic animals were not part of the ordinary diet, which instead drew protein from wild game, birds and reptiles of all kinds. In case of serious illness, however, chicken and pigs may be used for sacrifice. Despite contemporary observers describing the Co as well-fed, muscular, and not hard pressed for food and shelter (Mole 1970, pp. 179, 189), accounts from the wartime tell of Co longhouse communities fraught with disease of every imaginable kind (the crammed living space of the longhouse in some respects increasing vulnerability). Malaria was endemic; three types of typhus were present, and other diseases listed cholera, typhoid, yaws, leprosy, tuberculosis, venereal diseases, periodontal diseases and smallpox. Other plagues stemmed from parasitic and fungus infections, some presumably related to certain taboos against washing, and nutritional diseases related to deficiency of a range of vitamins and minerals (Schrock et al. 1966, p. 11; USASWS 1965, p. 11). These were invariably linked to the workings of demons or angry spirits, and attempted redressed by rituals and animal sacrifices. When the chief and elders decided on the periodic move of the longhouse, it could well be reasoned upon evil spirits taking possession of the house and its inhabitants. A missionary visiting the neighboring Co-tu around 1960 described how she offered malaria medicine to cure a sick woman, but was met with agitated refusal from her family: 'She's sick because she has no pig for a sacrifice to the demons. Give us a pig to sacrifice for her and she'll be well!' (Smith 1965, p. 25). The overwhelming trust in spirits as opposed to material factors and hygiene nourished the Vietnamese prejudices of the Co as a 'backward' people.[16]

5 The Forest Environment

The Co used to live in a rich forest environment with interchanging patches of highland fields, secondary forest and old primary forest, many places developing dense, impenetrable jungle. They were renowned for their ability to navigate in the jungle and to read the tracks of any moving animal or human. Both their identities and vast indigenous knowledge were intimately connected to the forest, which they both cherished for its wealth and respected for its spiritual forces and constant challenges. In its pre-modern state, the forest would protect itself, so to speak, as its natural inhabitants, large and small, combined with the many other hazards of the forest would hold humans at bay.

Most feared were the tigers that roamed the mountains, frequently carrying off children and old people from around the longhouses, as well as attacking hunters in the jungle and people working in the fields. No one would move about alone outside the longhouse community, and many elderly Co people speak of a general sense of fear. When living in the forest Co people were of rather small stature (adults weighing only 40–60 kg) in comparison with male tigers weighing up to 180 kg. Mr. Khoi, now in his 60 s, tells of how several family members were caught and eaten by hungry tigers as was his uncle when climbing a tree for honey. Once, three people from his longhouse were killed in a single day. Tigers were described as a constant threat to highland people as late as the 1960s (Smith 1965, p. 103). Elephants were equally feared, though mostly as a threat to crops. Fields many places needed constant protection with fences or rows of pointed bamboo sticks to ward off animals. Still quite commonly, however, crops were destroyed by marauding wild elephants and pigs, or by monkeys that would take their pick of fruits and vegetables.

Hunting, trapping and fishing provided the daily intake of animal protein. Fish up to 30–40 cm were caught in rivers and streams along with a variety of tortoises,

[16] Modern medicine did not appear to reach the highlands in this period. French era medical programs were discontinued (Hickey 1982b, p. 57). Even though tribal representatives desired medical support and the Vietnamese government had a policy to that effect, as one source states, 'active social work is hindered by a dispute concerning responsibility for administration' (US Army 1966, p. 28).

crabs and snails. Using crossbows and spears, the Co hunted for dear and boar, often collectively by driving them into pointed bamboo sticks to kill or stall them. But like other hunter-gatherers they caught smaller animals like monkeys, rodents, reptiles and a range of birds without preferences. Particularly those animals seeking food in the cleared fields were common game. Smaller game might be caught in inventive traps and snares, which still can be seen in the mountains today.

Yet the dangers and unpredictability of life in the forest did not thwart the Co people's emotional attachment to it. They express a genuine appreciation of its beauty, colors and richness and an affiliation with its wealth of animated objects and places; this remains strong among the elderly Co people today. Their knowledge of the forest habitat and ability to interpret the fertility of mountain fields meant that their activities inflicted only minimal disturbance (e.g. Tran 2008). Co people emphasize that despite their use of fire for clearing forest fields, which has been constantly criticized as an unsustainable practice and a source of deforestation, they always used it in a sensitive way and closely managed it to protect the surrounding forest. As much of the larger growth was felled by the men in the autumn as a first preparation of a new field, and was left to dry for several months before being burned in winter, there was always a clear demarcation of the field to be burned. Apart from the use of fire, traditional Co technology had only limited capacity to interfere with the forest ecosystem, which until the 1960s presumably had remained in a state of dynamic balance between forest and fields for many centuries. Only during the American war did the Co acquire guns (one of the three modern tools of guns, trucks and chainsaws which had the most dramatic impact on the forest ecosystem).

In the context of the present attention to natural disasters such as typhoons, floods and landslides, elderly Co people memorize a highly resilient forest environment before the modern era. Typhoons, a common threat to lowland communities throughout history, were rarely hazardous in the mountains because the old natural forest and dense secondary forest offered sturdy protection. Similarly, landslides and soil erosion, which are now huge problems on bare or acacia clad hillsides (see Chap. 1), are not remembered to have occurred on any large scale. An informant stated that the Co would never build their longhouses on too steep a ground for fear of landslides, indicating that they were a known threat. Another elderly informant said that once 'the earth was moving and cracking open, with steam coming out', presumably referring to an earthquake or volcanic activity. Only flash floods appear to have been a serious threat in the Co lands. After heavy and continuous rains, water masses would pour down the gorges, uproot trees, flood the lower fields and sometimes cause loss of life.

6 The Post-War Sedentarization Process

In 1975, when the North Vietnamese forces achieved full control of lowland Vietnam, the Co like the general population of Quang Nam had been through unspeakable suffering. I addition they had been uprooted from their culture and traditional way of life. The formal termination of the war did not at once bring peace to the highlands. Fighting continued for several years, and many highland minority groups paid a high price for their loyalty to the Americans.[17] As part of programs for New Economic Zones and Resettlement for Sedentary Cultivation immediately after the end of the war (ADB 2002, p. 10), all Co people were moved out of the mountains and resettled in new lowland villages in Quang Ngai province to the south and Quang Nam province to the north. This was premised on a general notion among Vietnamese scientists that swidden cultivation represented a poor and backward economy (Nguyen 2011, p. 46). In Quang Nam, the Co were concentrated in several villages along the road from Highway 1 on the coast towards Tra My, mainly in Tam Tra and Tra Dong communes. No exact information is available on the composition of these villages, but in several cases, presumably where former longhouse chiefs were already loyal to the North, an entire longhouse community was moved into a new village or hamlet (sub-village). Thus, Lê Văn Phúc (mentioned above), having fought for the North, became leader of Tam Tra's village 6, where he presently lives in an old wooden house.

On the question of resistance to sedentarization, people will merely say that they 'followed the guidance of the authorities'. After years of suffering, including destruction of their houses and widespread famine in Quang Nam in the final years of the war,[18] Co communities were already splintered. Under the control of the Department of Fixed Cultivation and Sedentarization each village was allocated land for paddy cultivation, of which much had to be cleared and dikes be built. Co people no longer had legal access to forest products and experienced great difficulties in producing enough food.

With the highland depopulated, the forest became a state asset. Indiscriminate exploitation of forest resources by state cooperatives and institutions soon began (Vien 2012, p.1). The State Forestry Company expanded from the North and within a short time span huge forest resources were depleted (Vien 2012, p. 2, see also " The Push for Plantations: Drivers, Rationales and Social Vulnerability in Quảng Nam Province, Vietnam"). Japanese investors soon became involved in new coffee plantations on former Co lands and rubber plantations followed. Wildlife disappeared rapidly; elephants had roamed the Co lands until around the end of the war, when the last was shot and the meat distributed between villages in Tam Tra.

The shortcomings of the state forest management system, such as evident in the dramatic expansion of barren lands,[19] and continued conflicts with local communities, which refused to be cut off from highland fields and forest resources, induced the state to begin allocating forest land to cooperatives, communes and households

[17] According to some sources, more minority people were killed in continued fighting after 1975 than during the war itself (e.g. Wiesner 1988).

[18] A famine developed in 1971, on top of which a government ban on the private sale, storage and transport of rice was introduced in 1973 to block the enemy's access to food.

[19] Some sources suggest that the rate of barren lands expanded from 10 % of Vietnam's land area in 1943 to 40 % in 1995 (Vien 2012: 2).

in 1983. By the late 1980s, most of this land was effectively allocated or reallocated to private households. Resolution 10 of 1988 thus instructed Party and government to allocate forest land to local users to 'protect, care, plant, exploit and take advantage of forest products…' to improve their livelihoods (repeated in the Law on the Protection and Development of Forests in 1991). Households were granted long-term user rights and access to support and loans. At the same time, a number of programs were implemented to counteract the returning practice of swidden cultivation in the highlands (such as Program 135), mostly focusing on state support to reforestation by means of plantation forestry.

Forests and natural resources in principle remain under state control in Vietnam. The Land Law of 1993 stipulates that land is the property of the entire people, uniformly managed by the state. Yet this law allows users to exchange, transfer, lease, inherit and mortgage land use rights (Vien 2012, p. 5). Since the early 1990s, a long list of Decisions and Decrees were issued to regulate the use of forest land. The 1992 327 Program (Decision 327/CP) had a particular bearing on minority populations as it was a large scale program to 'improve forestry productivity' by massive forest planting activities in the highlands which was to be achieved by using cultivation technology developed in the lowlands (Vien 2012, p. 7). Acacia was the main tree crop due to its rapid growth, but had a secondary aim of achieving fixed cultivation and permanent settlement. The 661 Program of 1998 (Decision 661/QD-TTg) had similar objectives, namely to improve both the quality of forest and the livelihoods and job opportunities in the highlands, while at the same time sedentarize highland people. The outcome of these programs have been much debated in Vietnam, and in addition to rampant corruption and misappropriation of funds (e.g. Vien 2012, pp. 8–10; ADB 2002), it has been pointed out that both forest quality and biodiversity have been steadily degraded despite increasing forest cover.[20]

Today the elderly Co people, who once lived in the forest longhouse communities, will say that there is little 'real forest' left, except for small fragments of high growth on inaccessible hilltops. They complain about deforestation and its consequences. As they see it, the soil is drying out to become hard and infertile, and the weather is feeling hotter without the shelter and moisture from the forest.[21] In the mountains now located inside the Phu Ninh Protected Area[22] which was formerly part of the central Co lands, there are 15,000 ha of land classified as 'protected forest'. But it is increasingly used for private acacia plantation forestry, mixed with some traditional Co highland fields and secondary forest or shrubs. According to locals (and confirmed by personal observation), what remains of the

---

[20] Both statistics and estimates are unreliable. One source mentions that in 1995 merely 5.5 % of the remaining Vietnamese forests may be considered 'rich forest' (FAO 2001, p. 186).

[21] Deforestation will affect soil quality and can actually affect local weather significantly (e.g. Butler 2012).

[22] Phu Ninh Protected Area covers an area of 23,409 ha, of which 27.4 % are in Phu Ninh district and 72.6 % in Nui Thanh district. It was designated in 1986 as part of the Phu Ninh Lake project for irrigation of lowland areas, water supply for industry and domestic use in the Tam Ky area, and electricity generation.

forest is rapidly disappearing in illegal logging, an apparently unstoppable activity across the province.

Many local Kinh people and officials tend to blame the Co people for forest destruction, but according to themselves it is primarily Kinh people felling the forest for money and then burning it in order to blame the Co. Whatever the truth is, it remains a fact that trucks and expensive equipment as well as the further transportation and sale of hardwood are entirely dominated by Kinh people.

A separate issue relates to the 1993 Land Law which as an instrument of the Doi Moi economic reform allocated land to individual peasant households. Under this law, land users are required to cultivate the land intensively and continuously. Moreover, reclamation of waste and fallow land is encouraged and state institutions have the right to reclaim 'abandoned land' for instance by allocating it to private individuals. The problem for traditional highland agriculturalists is that mountain land not covered by high forest has commonly been classified as 'unused land' by the authorities (Kemf and Quy 1999, p. 166), even if it is part of a shifting agriculture cropping cycle. The implications for the Co people are that private or corporate plantation forestry has gradually encroached upon mountain land and initiated a new struggle for land (Red Book) certificates among both wealthy locals and outside investors. Co people commonly claim that the Kinh can influence local authorities to take sides with them, either through ethnic loyalties or money. As a consequence, the Co face reduced living areas and lack of protection from especially commune level authorities (see also ADB 2002, p. 18). When Co people gradually gained access to resuming their highland agriculture in recent decades, it was preconditioned on private land use right, as opposed to the traditional collective rights. It was also in competition with lowland Kinh people who have benefited disproportionately from better financial resources and better access to local government. A process of social and ethnic differentiation was set in motion. The closer we get to the present time, the larger the land holdings of individual people. In fact they tend to grow year by year. Local Kinh informants reported that several wealthy Kinh people now control forest land in excess of 60 ha inside the Phu Ninh Protected Area, acquired by means of bribery to the local commune administration and possibly by means of Co people acting as 'straw men' and subsequently transferring the land rights. In Vietnam as elsewhere in Southeast Asia, the present quest for land is symptomatic of a general process, where previously neglected territories have become the object of a new gold-rush (Scott 2009, p. 11; IWGIA 2012).

Access to land certificates (Red Book registration) is now a crucial issue for everyone. The household survey indicated that lack of registration was a general problem for Co households—particularly their highland forests and fields lack registration. In fact, of the 17 households included in… (from Nui Thanh/Tam Tra, village 6 and Bac Tra My/Tra Giang, village 3) only three households had registration for all their land. For nearly all… what, their highlands lacked registration, according to the Co this is because the municipality officials thought it inconvenient to walk several hours into the mountains to measure the land for the records. Both bribery and ethnic loyalties in connection with common land registration were also reported, disfavoring the poorer minority people.

Poverty looms large among the ethnic minorities of Vietnam, unable to catch up with Kinh and Chinese population groups (Swinkels and Turk 2006; VASS 2011), and data from Quang Nam only confirm this trend (ADB 2002, p. 17). No less that 13 of 17 Co households in the survey were classified as poor by local authority (76 %), as compared to approximately 25 % of the Kinh. Comparing yearly incomes was not easy because the Co people still have a much higher degree of subsistence farming on highland fields and mixed livelihoods. However, a rough calculation would show that Co incomes on average were only half of that for the entire group. There were several instances of large families (up to 8 members) with extremely low incomes, estimated at less than $50 per month.

7 Cultural Identity, Education and the Young

The generation of Co people born after 1975, and their children in the next generation, have grown up in Vietnamese style villages with permanent housing of wood or concrete. One obvious difference from Kinh villages, however, is that houses tend to be small and many are still built of wood. The village 6 of Tam Tra has a community hall built with support from international donors and an open space for festivals or for kids to play football. Other differences show only indirectly. In terms of education, material from the general survey (see "Climate Change, Adaptation and the Environment in Central Vietnam") indicates a lower level of average schooling, and particularly many young Co boys leave school prematurely. Two young sisters, Nguyen Thi Ngoc Loan and Nguyen Thi Yu who run a small shop in their parent's house in village 6, tell of their experiences as Co children in the local school. They felt discriminated against, as they were constantly reminded of their inferior status as Co, made to kneel before the teacher and were punished much harder than Kinh children. They claim that Co children generally are given lower marks and more easily evicted from the school, contributing to a common inferiority complex among minority people (ADB 2002, p. 19, see also "Rural Households: Socio-Economic Characteristics, Community Organizing and Adaptation Abilities" on schools). Some improvement has occurred in recent years because several Co people have educated themselves to become teachers in their former school, offering better conditions for children of their own community.

Local schools exclusively teach in Vietnamese language already from the first grade (local kindergartens may be to bilingual), and the Co language is not accepted in class. It may only be spoken in the schoolyard and after school. Co children are still mostly bilingual, and in the villages a mixture of languages is used when they play, but many parents now teach their children Vietnamese before the Co language in order to give them a better chance in life in regards to education, trade and employment. The two sisters named above, of which the elder has a small baby, tend to use Vietnamese more often now and will teach Vietnamese as a first language to their own children, yet, as they say in an amiable tone, 'We remember the discrimination we suffered and we will fight back when we grow up!' The interaction between young Kinh and Co people is not without friction, and oftentimes fights break out over issues of land or irrigation water. In 2011 a fight between several young from either side ended in a Kinh man throwing acid in the face of a Co man and police were called in. These are social conditions that impact the reproduction of Co ethnic identities in contemporary Vietnam. Many young Co people feel marginalized and discriminated against in terms of access to employment and land certificates. They particularly blame commune authorities for siding with Kinh people, while having more faith in district and provincial authorities.

As opposed to the wholly modernistic identities that prevail in Vietnamese society, the Co identity remains attached to the forest even today. Those Co men possessing forest land will go to the forest every day, often walking several hours, and stay there during the day to work in fields and acacia plantations. The women will follow when their labour is needed such as to plant, weed, harvest, prepare the acacia trunks for sale or to carry huge bundles of firewood back to the village. Even some of the young children will spend all day herding cattle into the mountains to graze, having left school at a very early age. As one such child said, 'I left in the third grade because I was not smart enough for the school'. They will all express appreciation of the forest, however, and a commitment to its spiritual order. The young girls Loan and Yu, for instance, passionately describe the remains of old orchards in the mountains, where they may go and feast on oranges, mandarin, pineapple, guava and other fruits, 'It is like going to Heaven', they say. But the place is believed to be sacred and protected by spirits, and thus surrounded by taboos 'You will die if you bring the fruit out of the forest!' At the same time they point out, 'We still honor our ancestors the Vietnamese way!'

Other strong identity markers are the annual Co festival, held in October, and the old material culture comprising gongs, dresses, jewelry, jars and weaponry that used to be taken out at these events. Many items have become too fragile of age. Animals are no longer sacrificed, but the women will perform the traditional singing and dancing acts, each one displaying her own personal style while singing old songs and beating the gong.

8 Conclusion

For several decades, Vietnamese minority policy has had strong wartime reminiscences in contrast to which the present government may be said to 'seal the past and look to the future'. Yet the two divergent views of highland development that became apparent after the Doi Moi reforms (Rambo et al. 1995, p. 21) hold sway even today. One view, being dominant in official discourse and policy making, sees the highlands as under populated areas with immense pools of natural resources, constituting a new frontier for national development. An alternative view, more prominent among researchers, sees the highlands as fragile ecosystems, already overpopulated and suffering from progressive environmental degradation and loss of biodiversity, with indigenous population groups locked into poverty as they lose out to progressing state and private exploitation of local resources. Seen from the latter perspective, the main objective of development is to undo past mistakes, restore degraded habitats, and give renewed hope to ethnic minorities.

There is still agreement among those individuals old enough to have experienced traditional Co life in the forest that life is easier and more comfortable today, and certainly much better than in the war times that fill the memory of the elder generation. Malnutrition, disease and discomfort were part of forest life, and the nature of exchange relations with lowlanders often added to their insecurity. In recent years, acacia plantation forestry in particular has brought rising incomes for many Co households, yet not on a scale as seen among the most successful Kinh people. Both poverty and difficult access to land certificates, legalization of ownership to their traditional lands, remain key problems and add to their vulnerability in several respects. Firstly, after disasters like typhoons which destroy acacia plantations they may be less likely to receive compensation, and secondly, without legal land rights they will generally have less protection against possible land grabbing as much as they are in a weak negotiating position in case of hydropower or other construction projects on their lands. Similar to the situation for other minority people in Quang Nam and in general, they have higher than average incidences of poverty, and large population segments placed between the Vietnamese and the international poverty levels, inevitably making them vulnerable to a range of stresses, including climate change.

The modern history of the Co people is intimately linked to that of their natural environment, the forest. Historically controlling vast and precious resources, yet never profiting on them other than for their simple livelihoods, they were caught in the sweeping advance of Vietnamese-led modernization. The gradual but consistent inclusion of their mountain-forest environment in warfare, political tactics, resource exploitation and lowland population resettlement has impacted their way of life as much as direct integration policies. The forest itself, which preciously formed a rich and dynamic environment, is rapidly being converted to mono-cropping plantations interchanging with barren hills, while subjected to landslides, soil erosion and diminishing biodiversity. This is a result of human intervention and unconnected to climate change, towards which the present mountain-plantation environment presumably is increasingly vulnerable.

The retreat of the natural forest in this region can also be said to mark the termination of a historical battle between highland and lowland Vietnamese. But as much as the transformation of the natural landscape continues today, so does the transformation of the Co identity. In the final account, the Co people's access to modernity has been traded for shattered identities, cultural disintegration and possibly eventual assimilation. Against the background of Vietnamese government policies and public attitudes to ethnic minorities, which may have improved over time but still uphold strong ethnocentric biases, there is still fuel for contention. The question is how young Co people will make sense of the transformation of their culture and environment.

References

ADB (Asian Development Bank) (2002) Indigenous Peoples/Ethnic Minorities and Poverty Reduction, Vietnam. Manila, ADB

Bruun O (2006) Precious Steppe: Mongolian nomadic pastoralists in pursuit of the market. Lexington Books, Lanham, MD

Butler R (2012) Impact of deforestation: local and global consequences. http://rainforests.mongabay.com/0902.htm

Chu TS (2004) Nguoi Co. Hanoi, Nha Xuat Ban Tre

Condominas G (1951) Aspects of a minority problem in Indochina. Pac Aff 24(1):77–82

Condominas G (1977) We have eaten the forest: the story of a Montagnard village in the Central Highlands of Vietnam. Hill and Wang, New York

Dang NV, Chu TS, Luu H (2000) Ethnic minorities in Vietnam. The Gioi Publishers, Hanoi

EVC (Encyclopedia of Vietnamese Culture) (2012) (www.vhv.vn/vhv_nguoi-co). Accessed 10 Oct 2012

FAO (2001) Forests out of bounds: impacts and effectiveness of logging bans in natural forests in Asia-Pacific. Asia-Pacific Forestry Commission, Bangkok

Fall BB (1963) The Two Vietnams. Frederick A. Praeger, New York

Fall BB (1966) Vietnam Witness 1953–1966. Frederick A. Praeger, New York

Hickey GC (1970) Accommodation and coalition in South Vietnam. The RAND Corporation, Santa Monica

Hickey GC (1982a) Sons of the Mountains: Ethnohistory of the Vietnamese Central Highlands to 1954. Yale University Press, New Haven (Reproduced in Hickey 2003b)

Hickey GC (1982b) Free in the Forest: Ethnohistory of the Vietnamese Central Highlands, 1954–1976. Yale University Press, New Haven (Reproduced in Hickey 2003b)

Hickey GC (2003a) Ethnographic notes: Cua. In: Hickey 2003b

Hickey GC (2003b) Glints of a shining World: a Gerald cannon Hickey reader. Radix Press, Houston, p 2003

IWGIA (International Work Group for Indigenous People) (2012) Update 2011: Vietnam. IWGIA Year book. Copenhagen, IWGIA

Kelly FJ (1973) History of special forces in Vietnam, 1961–1971. Department of the Army, Washington, DC

Kemf E, Vo Q (1999) Ethnic minorities and protected areas in Vietnam: The effect of land use on biodiversity in the buffer and core zones of Yok Don National Park. In: Colchester M, Erni C (eds) Indigenous peoples and protected areas in South and Southeast Asia. Copenhagen: IWGIA Document 97

Le CD (2004) The central highlands: a North Vietnamese journal of life on the Ho Chi Minh trail, 1965–1973. The Gioi Publishers, Hanoi

Le VH et al (1993) Dai Viet Su Ky Toan Thu 1272–1697 (The complete history of Dai Viet, 1272–1697). The Gioi Publishers, Hanoi

Leach E (1954) Political Systems of Highland Burma. Cambridge University Press, Cambridge

Masako I (1999) Vietnamese language policies towards ethnic minorities—the case of the Tay-Nung language. Kotoba to Shakai (Lang Soc) 2:81–96

Mole RL (1970) The Montagnards of South Vietnam: a study of Nine Tribes. Charles E. Tuttle Company, Tokyo

Nguyen VC (2011) From Swidden cultivation to fixed farming. In: Taylor P (ed) Minorities at large: new approaches to Minority Ethnicity in Vietnam. ISEAS, Singapore

Phillips RL (1962) Here are the Tribes. Jungle Frontiers XVI:12–13

Radix Press (2002) Minority groups in the Republic of Vietnam. Radix Press, Houston, TX

Rambo T et al (eds) (1995) The challenges of highland development in Vietnam. Honolulu, East-West Center, Program on Environment

Raskin MG, Fall BB (1965) The Vietnam reader: articles and documents on American Foreign policy and the Vietnam crisis. Random House, New York

Salemink O (2003) The ethnography of Vietnam's central highlanders: A historical contextualization. 1850–1900. University of Hawaii Press, Honolulu
Schrock JL et al (1966) Chapter 3: The Cua'. Minority Groups in the Republic of Vietnam. Cultural Information Center, American University, Washington
Scott J (2009) The art of not being governed: an Anarchist history of Upland Southeast Asia. Yale University Press, New Haven
Smith LI (1965) Victory in Vietnam. Zondervan Publishing, Grand Rapids, MI
Swinkels, Rob and Carrie Turk (2006) Explaining ethnic minority poverty in Vietnam: A summary of recent trends and current challenges. World Bank, Vietnam: Draft paper for CEM/MPI meeting on Ethnic Minority Poverty, Hanoi
Tanham George (1966) War without Guns: American Civilians in Rural Vietnam. Praeger, New York
Tran SH (2008) Indigenous Knowledge of Thuong Ha People in their Cultivation on Northern Upland Vietnam. National Institute for Soils and Fertilizers, Hanoi
US Army, Headquarters, Field Force Vietnam (1966) Orientation on the Montagnards and related US/GVN policies. Reproduced in Radix Press 2002
USASWS (U.S. Army Special Warfare School) (1965) Montagnard Tribal Groups of the Republic of Vietnam. Fort Bragg, N.C
VASS (Vietnamese Academy of Social Sciences) (2011) Poverty reduction in Vietnam: achievements and challenges. Hanoi
VCDS (Vietnamese Central Department of Statistics) (2009) General investigation on population and housing in 2009. Hanoi
Vien TD (2012) Forestland management policies in vietnam. http://www.cares.org.vn/webplus/Article/Forestland%20management%20policies%20in%20VietNam%20An%20Overview.pdf (Accessed 1 Nov 2012)
Vinh L (1965) The so-called movement for autonomy: FULRO. Pleiku, Banmethuot. Reproduced in Radix Press 2002
Vu H (1992) Collection of traditional Co stories. Da Nang Publishing House, Danang
Wiesner L (1988) Victims and survivors: displaced persons and other War Victims in Vietnam, 1954–1975. Greenwood Press

Viet Nam's Food Security: A Castle of Cards in the Winds of Climate Change

François Fortier

Abstract Since the 1980s, Viet Nam has achieved rapid economic growth and greatly increased food production and security. These results are based, however, on a model of industrial agriculture that has inherent social and environmental limitations and increasingly faces the structural constraints of climate change. This article questions industrial agriculture in general, and through the case of Viet Nam, its ability to sustain outputs and food security through the emerging crisis. It argues that while agro-industrial technologies and commodification are making the country particularly vulnerable to the imprecise and shifting context of a multifaceted crisis, the dominant response of the green economy, in Viet Nam as elsewhere, rests on unsubstantiated technological and institutional assumptions. Unchanged, such strategy will most likely lead to the collapse of Viet Namese agricultural production and a surge of food insecurity. In such a strategic vacuum, the article explores how agroecology offers a viable alternative, in parallel with the organization of production, distribution, and consumption through principles of food sovereignty.

Keywords Economic growth • Food insecurity • Food sovereignty • Agroecology • Climate change

1 Introduction

Since the launch of the Viet Namese *Đổi mới* (reform) process in the 1980s under the label of a *socialist-oriented market economy*, the country has achieved rapid economic growth and increased food production severalfold. The development of agriculture has had significant impacts on national food security, while providing livelihoods to 60 % of the population and generating about a quarter of Viet Nam's gross domestic product (GDP) (Carew-Reid 2008, p. 6). This gives, however, a false sense of security, for those achievements are entirely rested on a model of industrial agriculture that has inherent social tensions and mounting

F. Fortier (✉)
Centre d'études de l'Asie de l'Est (CETASE), University of Montréal, Montreal, Canada
e-mail: ttff21@gmail.com

O. Bruun and T. Casse (eds.), *On the Frontiers of Climate and Environmental Change*, Environmental Science and Engineering, DOI: 10.1007/978-3-642-35804-3_13,

contradictions, and which is also now facing structural challenges beyond its immediate organization of production, particularly those of climatic instability.

The point of departure of this article is therefore to question the viability of industrial agriculture, in general and through the case of Viet Nam, and the unlikelihood of it being able to sustain outputs and food security through this emerging crisis. This paper argues that, in the context of agricultural modernization, Viet Nam's dependence on technology and commodification is making it particularly vulnerable to the imprecise and shifting context of climate change. The strategic response to this challenge, branded as a *green economy* in Viet Nam as almost everywhere else, has been to provide technological and institutional remedies that, it is claimed, will decarbonize the economy while intensifying production outputs and maintaining food security amidst a changing climate. However, there is mounting evidence that such fixes will not ensure sustained economic activities, including industrial agriculture and food security. The magnitude of greenhouse gases (GHG) mitigation required and the time frame within which this is needed make virtually impossible the timely development and adoption of innovations that would be necessary but still remain, for the most part, unproven concepts. The article discusses how such strategy would most likely lead to the collapse of Viet Nam's industrial agriculture and the loss of its food sufficiency and security. The situation begs to urgently adopt agroecology, both as a means of ensuring food security and as a key contribution to GHG mitigation, and to restructure production, distribution, and consumption under principles of food sovereignty.

To make this argument, the article first reviews the process of agricultural modernization that has given rise to the current basis of Viet Namese food security, and the emerging biophysical context of climate uncertainty that now threatens agriculture. This leads to considering the response now being articulated by the Viet Namese state, well aligned on global climate governance trends. The article then turns to a critique of this response, examining the unrealistic expectations of hypothetical technological fixes that would address both mitigation of GHG emissions and adaptation of industrial food systems. The critique underlines the fragility of the current model, and its likelihood of collapse under multiple stressors. Beyond the shortcomings and deception of that model, the article concludes that only agroecology, and its organization under principles of food sovereignty, can offer viable alternatives.

2 The Rise of Productivity

From the mid-twentieth century, successive Viet Namese governments made significant investments in the modernization of agriculture. While it started slowly due to the protracted French and American wars, by the 1960s, this modernization had intensified in both North and South Viet Nam, notably with large-scale irrigation and drainage works, some level of mechanization, the introduction of modern high-yielding varieties, and the increasing application of agrochemicals (Taylor 2007, p. 10; Asian Disaster Preparedness Center 2003, pp. 15–17).

In the North, the government of the Democratic Republic of Viet Nam was an unwavering believer in the potential of human ingenuity, both in technological and organizational terms. It prioritized agricultural investments, notably in water management, resulting in the increase of irrigated areas of the total agricultural land from 42 % in 1955 to 64 % in 1960 (Bhaduri and Rahman 1982, p. 42). In the first half of the 1960s, the budget allocated to the agricultural sector had increased fivefold compared to the late 1950s, bringing an additional two hundred thousand hectares into production (Ha Vinh 1997, pp. 104–116). The state also promoted the use of short-cycle modern high-yielding varieties, enabling multiple production seasons for rice and other crops (Wiegersma 1988, p. 167). Policies were, however, more reserved with respect to agricultural mechanization. It adopted instead an approach of *technical duality*, discouraging the use of equipment that could displace large amount of labor while promoting technologies that had less such effect. "The net result was a sparse spread of small farm equipment. In 1977, the northern region had 10,160 tractors, which provided mechanized land preparation for 16 % of the area" (Pingali et al. 1997, p. 353). Agrochemical inputs were also scarce, for not being produced domestically and their importation constrained by limited financial resources. By the end of the 1970s, however, agrochemical inputs, mostly imported from the Soviet Union, had become more commonly used in the Democratic Republic of Viet Nam (DRV) (Fforde and Sénèque 1994, p. 21).

In the southern Republic of Viet Nam, the process of agricultural modernization was more intensive. Capitalist accumulation had already spread faster there during French colonization (Marquis 2000, p. 91), and production was largely commercialized by the 1950s. From then until 1975 and with American funds, the South provided testing grounds for the Modernization Theory by its intellectual father Walt Rostow himself (Pearce 2001). "The growth in the use of capital inputs in agriculture [...] was produced by a massive import of farm equipment, fertilisers and oil which was made possible by US aid money, by an enormous infusion of liquid capital into the hands of the landlords through the so-called 'Land to the Tiller' programme and by substantial loans supplied by the American subsidized Rural Development Bank" (Ngo 1984, p. 286). In 1975, high-yielding varieties of rice were planted on between 0.6 million hectares (Pingali et al. 1997, p. 351) and 1 million hectares of paddy fields, and accounted for about 30 % of the total paddy output (Young et al. 2002, pp. 10, 14). Between 30 and 40 % of the land in 1977 was being tilled by tractors (Pingali et al. 1997, p. 353), while large-scale water-control projects were drawn to replicate the hydraulic engineering of the Tennessee valley across the Mekong Delta (Käkönen 2009, p. 206).

By the end of the American war and in the aftermath of the 1975 reunification, the northern cooperative system slid deeper into crisis, while efforts to collectivize the south failed (Ngo 1993, pp. 169–173). In 1979, the government started experimenting with economic reforms, which culminated in the official launch of the *Đổi mới* (renovation) policies in 1986. While transforming the organization and relations of production by steering away from collectivization, agricultural market incentives were introduced, upstream and downstream channels were liberalized, and steps were taken to rejoin the global economy

(Kerkvliet 1995). The reform reiterated modernizing objectives by embracing the green revolution: it intensified irrigation and agrochemical use, adopted modern varieties, pushed the agricultural frontiers, and reclaimed costal lands. Within that period, Viet Nam's agricultural production grew remarkably. While the country produced 16 million tonnes of rice in 1986, the output reached over twice this figure by 2008, at 36 million tonnes. Productivity surged by 86 %, from 2.6 to 4.9 tonnes per hectare (International Rice Research Institute 2009, using FAO and USDA compilations). The largest share of that surplus comes from the Mekong Delta, supplying more than half the increase of the last 15 years.[1] As a result, Viet Nam went from being a net importer of rice in the 1980s to becoming the world's second largest exporter, selling 4–5 million tonnes per year—that is, 15–20 % of the globally traded annual volume (International Rice Research Institute 2007). The number of food insecure people was also reduced by one-third between 1990 and 2005. Beyond rice, other food outputs also increased, which contributed to some extent in reducing rural poverty. By official accounts, the share of the population living under the poverty line decreased from 75 % in 1988 to 15.5 % by 2006.

3 Looming Threats

Despite the productivity gains obtained through such an intensive modernization of agriculture, the actual magnitude of the benefits, the spread of their impact among different groups and regions, and their hidden social and environmental costs are not unchallenged (see notably Kolko 1997, pp. 104–116; Pincus and Sender 2008; Tran 2009, 2010). There has also been growing concerns with respect to national food production over the coming years, which could threaten whatever actual achievements on hunger and poverty. Such concerns are reflected in numerous debates within political, development, academic, and public spaces, and identify two major challenges to food production. One is the steady decline in cropping areas over the last decade, particularly paddy fields converted to other use. The other, still imprecise but even more worrisome, is that of climate change (see notably Ministry of Agriculture and Rural Development 2008; Government of Viet Nam 2009). Both are valid concerns, but the former issue is not discussed in this chapter, focusing instead on the viability of the Viet Namese model of industrial agriculture, and the unlikelihood of it being able to sustain current levels of outputs, and with it food security, in the emerging context of the climate crisis.

[1] National production in 1995 was 25 million tonnes, of which 12.8 million tonnes were from the Mekong Delta. Respective preliminary figures for 2008 were 38.7 million tonnes and 20.7 million tonnes. Hence: (20.7–12.8)/(38.7–25) = 57% (General Statistics Office of Viet Nam 2009b).

The Intergovernmental Panel on Climate Change (IPCC) lists Viet Nam among those countries to be most affected by climate change (Intergovernmental Panel on Climate Change 2007; see also Dasgupta et al. 2009; Institute of Strategy and Policy on Natural Resources and Environment 2009). Such vulnerability results from the combination of three factors: high exposure to natural elements, high sensitivity of socioeconomic structures to those elements, and low capacity to adapt by protecting structures or making them less sensitive (see notably Vogel and O'Brien 2004; Adger 2006; Eakin and Luers 2006; Füssel 2007; Nelson et al. 2007). Viet Nam's long coastlines, vast deltas, and location on the path of typhoons and monsoon rains mean that many parts of the country are widely exposed to weather extremes and sea-level rise (SLR) (Nguyen Huu Ninh, Vu Kien Trung, and Nguyen et al. 2007, pp. 2–3; Tran 2009; Carew-Reid 2008, p. 7; Chaudhry and Ruysschaert 2007, pp. 3–6; International Centre for Environmental Management 2009). The unpredictability of climate extremes will affect not only coastal areas but also the entire country through changes in temperatures and rainfalls, likely resulting in more droughts, floods, flash floods, landslides, and the spread of pests as well as plant, animal, and human pathogenic vectors. In turn, sea-level rise will lead to flooding and saline water intrusion, notably within aquifers. Climate change scenarios, even among the optimistic ones, expect severe SLR impacts for the Red River and Mekong Deltas. At a mid-range (B2) IPCC scenario, now optimistic by most accounts, a one-meter SLR is expected by or before the end of the twenty-first century. At such a level, 4 % of the Red River Delta would be submerged under seawater. For the Mekong Delta, however, it could be as much as 31 % (Carew-Reid 2008, pp. 14–15), as shown in Maps 1 and 2.

These new environmental attributes will severely affect key livelihoods in agriculture, forestry, fisheries, and aquaculture, as well as public health and infrastructure (Ministry of Natural Resources and Environment of Viet Nam 2008; Granich et al. 1993; Chaudhry and Ruysschaert 2007, pp. 5–6; Asian Disaster Preparedness Center 2003, pp. 19–21). For this reason, Viet Nam is also vulnerable because of the high sensitivity of its socioeconomic structure to the above biophysical impacts. The country's economy, and agriculture in particular, is very much dependent on climate, with about three quarters of the population living either in low-lying fertile plains potentially affected by SLR and fluvial floods such as the Red River and Mekong deltas, or in mountainous areas exposed to flash floods and droughts (Nguyen, Vu Kien Trung, and Nguyen et al. 2007, p. 3; Carew-Reid 2008, p. 6).

Mechanisms of climatic impact on agriculture are by now well documented, with numerous global models of changing patterns providing ranges of expected outcomes. Globally, those outcomes will depend on several factors: (a) international greenhouse gases mitigation policies that are or not implemented, (b) evolving humidity and temperature patterns, (c) the pace of sea-level rise with associated land losses and salinization, (d) changes in the prevalence and geography of plant and animal pests, and (e) the actual carbon dioxide fertilization that will stimulate or not the growth of certain crops. Even in the best-case scenarios, however, these factors are likely to affect very negatively the production, access to and utilization

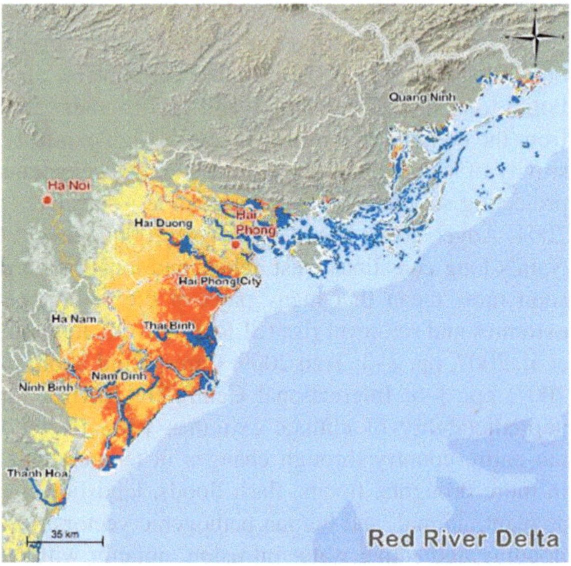

Map 1 Red river delta, inundated areas at SLR of 1 m (*blue*), 2 m (*red*), 3 m (*orange*), 4 m (*yellow*), and 5 m (*white*). *Source* Carew-Reid (2008), p. 18; reproduced with permission from ICEM, Hanoi 2011. Compass and scale added

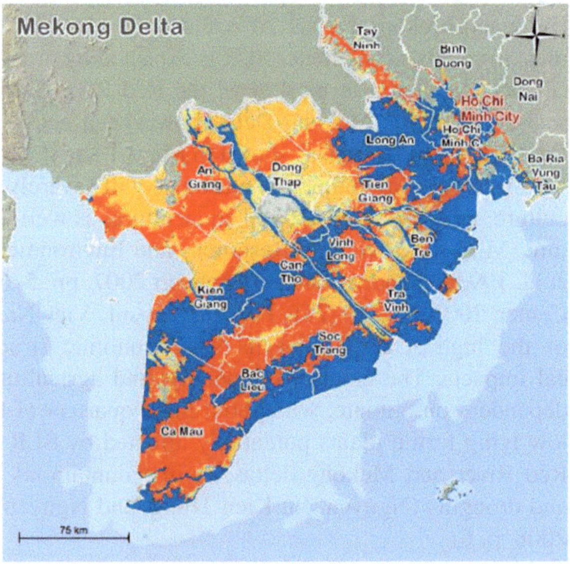

Map .2 Mekong delta, inundated areas at SLR of 1 m (*blue*), 2 m (*red*), 3 m (orange), 4 m (*yellow*), and 5 m (*white*). *Source* Carew-Reid (2008), p. 18; reproduced with permission from ICEM, Hanoi, 2011. Compass and scale added

of food. There will be a net loss of suitable land for agriculture at tropical latitudes for any rate of temperature increase, destabilization of supplies by extreme weather events, and possible market restrictions in view of increased risks (such as limiting exports to maintain strategic reserves). The nutritional value and safety of food is also likely to be affected by unfavorable growing conditions, warmer weather, and water cleanness (Turral et al. 2011; Batchelor et al. 2009; Bates et al. 2008; Food and Agriculture Organization of the United Nations 2008; Ericksen 2008;

Ericksen et al. 2011; Schmidhuber and Tubiello 2007; and specifically for the case of Vie Nam see Yu et al. 2010; Nguyen 2011). As a combined result of such threats, and simulating different variables, models applied to major grains (rice, wheat, maize, millet, sorghum) predict single and sometimes double-digit drop in production, and a 10–100 % surge in food prices relative to non-climate change scenario by 2050 (Nelson et al. 2009a, pp. 5–9; see also Cline 2007; Schmidhuber and Tubiello 2007, p. 19706; Rosegrant et al. 2008, p. 19; McIntyre et al. 2009, pp. 47–49; Battisti and Naylor 2009; Nellemann et al. 2009; Fedoroff et al. 2010, p. 833). Of particular relevance to Viet Nam, scenarios reviewed in Nelson et al. (2009b, pp. 4–5) indicate how irrigated rice will likely be severely affected by climate change, loosing up to 19 % in crop productivity, even before accounting for land-use change, salinization, or other abiotic stresses.

The third factor of Viet Nam's vulnerability is its adaptive capacity—that is, the ability "to better cope with, manage or adjust to some changing condition, stress, hazard, risk or opportunity" (Smit and Wandel 2006, p. 282). This capacity is constrained by the country's limited financial resources, reflected in a purchasing power parity of just under USD 3,000 per capita (International Monetary Fund 2010). In addition, poorer people within the country, especially poor women and youth, are often the most exposed and sensitive to climate change, involved for example in agricultural livelihoods on disaster-prone lands. Those groups are also the less endowed and empowered to recover from disruptions and for shifting to alternative livelihoods (Adger 1998, pp. 5, 10–11; OXFAM International 2008, p. 13; Chaudhry and Ruysschaert 2007, p. 2). It could be argued, in turn, that the Viet Namese society has a long history of adaptation to structural change, notably foreign occupations, wars, and natural disasters (Kelly and Adger 2000; Adger 1999, 2000), which should bode well for its ability to face emerging climatic threats. Yet, as will be discussed below, the economic reforms that have redefined the country's productive and social relations since the 1980s may also have inadvertently but severely weakened that adaptive capacity.

Viet Nam is therefore vulnerable to climate change for being exposed to natural elements, for being a society and an economy sensitive to that exposure, and for having limited means and decreasing options to adapt by mitigating impacts or reducing sensitivity. Beyond land conversion and climate change, however, it is also worth mentioning that the country's agriculture is facing other but largely ignored structural constraints, which could have significant synergetic impacts (see discussion on multiple stressors in O'Brien and Leichenko 2000; O'Brien et al. 2004; Eakin and Luers 2006; Adger and Barnett 2009; McIntyre et al. 2009, p. 50; Max-Neef 2010). First, industrial agriculture depends on the continuous availability of fossil fuels for both energy and agrochemical inputs, requiring "ten calories of fuel for every calorie of food produced—and this does not include the energy used in processing, packaging, and shipping the final product to market" (Philipps 2007, p. 97; see also Vandermeer et al. 2009; Martínez Alier 2011). Yet, like the rest of the world, Viet Nam is already facing the end of cheap fossil fuels, with production peaking and average prices rising for both energy and agrochemical inputs (Li 2007, pp. 452–463; IEA 2010; Hong Van 2007; Viet Nam Business

News 2010). In addition to this global fossil fuel crisis, Viet Nam is also under further energy constraints from problematic hydroelectric production due to repeated droughts over the last decade (Binh et al. 2010). As electrical power is supplied to priority urban and industrial areas, interruptions in the past few years have been particularly severe for the countryside. This has impacted most industrial agriculture, in which electrical power is extensively used, notably for irrigation and processing (Viet Nam Peasant Association 2010).

Second, Viet Nam's industrial agriculture also depends on a stable global trade regime for the country's realization of a large share of its high-value production. It currently imports more than one-third of its agrochemical inputs and raw materials to produce further inputs domestically, while three quarters of hybrid rice seeds are procured from China (Genetic Resources Action International 2008). Viet Nam also depends on global trade to commodify its production: it has exported nearly 80 % of its GDP in 2008 (General Statistics Office of Viet Nam 2009a), while the KOF Index of Globalization—measuring economic, social, and political indicators of global integration—nearly doubled since the start of *Đổi mới*, raising from 25 in 1987 to 48 in 2007 (KOF 2007). This exposure to global markets is already reflected in fluctuations of commodity prices and sporadic contractions of foreign markets (Greenfield 2004; Tran 2009a). As the neoliberal global trade regime shows signs of fatigue, the jury is out on whether or not this may lead to the shrinking of that paradigm in the coming years (see notably Panitch and Gindin 2009, pp. 21–29; Radice 2009; Chorev and Babb 2009; Gills 2010), but points toward a possible wave of re-regulation which would constrain global markets, and foreclose the opportunities that Viet Nam's modern agriculture has invested in.

4 Dominant Response

While little is said of the energy and trade stressors, the Viet Namese state has fully acknowledged and has been very concerned by how climate change may systematically disrupt its economy, and its agriculture and food security in particular. For many years now, government agencies have responded to the threat by building research and institutional capacity, and by devising policies in priority sectors. In 2008, the Ministry of Agriculture and Rural Development (MARD) adopted an Action Plan Framework for Adaptation to Climate Change in the Agriculture and Rural Development Sector for 2008–2020. The document raised numerous red flags with respect to agricultural production, food security, and natural disasters. In December 2008, the prime minister adopted the National Target Programme in Response to Climate Change (NTP-RCC) prepared by the Ministry of Natural Resources and Environment (MONRE). The NTP-RCC explicitly raises food security risks with particular attention to rice and fisheries (Government of Viet Nam 2008). By late 2009, the prime minister issued a resolution on national food security, questioning land conversion and calling for renewed efforts to increase agricultural productivity through expanded irrigation, other hydraulic engineering works, and agricultural research (Government of Viet Nam 2009). In practice, the government's response has indeed planned for better flood protection infrastructure, including a 2008 MARD proposal for USD 676 million investment in dike enhancement across the country (Biggs et al. 2009, p. 212), as well as modified crop cycles and technological solutions such as hybridization and genetic modification for increased tolerance to heat, drought, water logging, pest, or salinity (Ministry of Natural Resources and Environment of Viet Nam 2003, p. 66). Most recently, the Institute of Hydro-meteorological and Environmental Sciences has also revised upward its assessment of biophysical impacts, beyond the IPCC's mid-range B2 scenario. The authority expressed particular concerns with respect to agricultural livelihoods and production in the Mekong Delta (Viet Nam News 2011).

4.1 Assumptions About Mitigation

The strategic paradigm reflected by those policy documents is well in line with Viet Nam's own continuum of modernist aspirations, and also echoes the discourse and practice that drives global capitalism to its green economy redux through ecological modernization (Fortier 2010; and theoretical discussion in Foster 2008, pp. 536–540; Bäckstrand and Lövbrand 2007, pp. 126–131; Luke 1996, pp. 6–13; Brooks et al. 2009). It resonates with policies advocated by international development agencies and the agri-food industry that emphasize environmental control, biotech research, and technological intensification toward a *climate-smart* agriculture, and promote *climate-ready* policies based on better information gathering and sharing (see, for example, Asian Disaster Preparedness Center 2003, pp. 25–27; United Nations Convention on Climate Change 2006; Nelson et al. 2009a; Fedoroff et al. 2010; World Bank 2010, pp. 154–56; UK Government Office for Science 2011; and specifically for Viet Nam, Yu et al. 2010).

However, there is scant evidence that institutional development and technological innovations could in fact deliver the pace and type of fixes that would sustain current economic activities, including industrial agriculture. On the one hand, global GHG mitigation efforts are failing to provide any hope of climate stabilization, which Viet Nam would need as any other country to maintain its growth trajectory. Despite the accepted target of 2 °C increase in global average temperature above preindustrial levels (Smith et al. 2009; Richardson 2009), current voluntary abatement pledges from the Cancún Agreements are well off the mark, and not much more is expected from Durban as this article goes to press. Even if those pledges were implemented by 2020, followed by long-term draconian reductions, it would still leave emission gaps of 3–16 $GtCO_2e$ per year—that is, 7–36 % too high (United Nations Environment Programme 2011). Such emission pathways are well above what would give a minimum chance of keeping the global average temperature within the 2 °C target (Anderson and Bows 2008; Rogelj et al. 2011). Furthermore, that 2 °C threshold has long itself been contested for being overly optimistic about Earth system stability

(Hansen 2005, p. 277). In recent years, several key articles were published focussing on CO_2 concentration targets, highlighting how current levels (391 ppm CO_2,[2] and rising about 2 ppm per year) already have deleterious impacts on measurable long-term climatic trends. Conclusions are that safer targets for climatic stability should be at 350 ppm CO_2, possibly even 300 ppm, for a maximum rise in temperature of 1.5 °C (Hansen 2005, p. 275; Hansen et al. 2008, p. 226; Rockström et al. 2009a, b). The magnitude of greenhouse gases abatement required to stay within the not-so-safe limit of 2 °C, let alone 1.5 °C, and the time frame within which these are needed, would therefore require tremendous regulatory incentives and institutional support. Yet as seen above, the mitigation strategy of ecological modernization and its green economy falls well short of such an effort. The global governance model of the industrial society, despite two decades of explicitly confronting climate change, remains unable to face the cumulative stress that will eventually lead to dysfunctional production, including that of its agricultural model. While Viet Nam alone, through its relatively small contribution to past or even present GHG emissions, cannot take much blame for such trends, its current response does nothing either to prevent further environmental damage or foster realistic mitigative actions.

On the other hand, ecological modernization rests its hopes on the rapid development and deployment of socio-technical systems (technologies with institutions, knowledge, and relations of power), which assumes that capacity will be forthcoming to (a) drastically reduce GHG emissions through much lower material-energy throughput, measured as the carbon intensity of wealth created (Pacala and Socolow 2004), and (b) reabsorb GHG in excess of "overshooting" thresholds before points of no return are passed in climatic disequilibrium, notably through land-use management (Bäckstrand and Lövbrand 2006). It then becomes essential to assess whether the proposed technological response can indeed reasonably be expected to quickly and thoroughly decarbonize economic activities, provide significant GHG sequestration, or eventually geo-engineer the climate if the pace and scope of those other outcomes fall short.

On close examination, however, those assumptions are unsubstantiated for a number of reasons. Like mitigation targets, they too are unrealistic, turning the hopes of the green economy into a misguiding mirage. For one, the actual carbon intensity reduction of the past two decades has been of 0.7 % per year. Yet, even if global emissions were to peak in 2015, and accounting for population and economic growth, the world economy would need to reduce that carbon intensity by 7 % a year—that is, ten times faster than it actually is—to have a reasonable chance of not rising average temperature by more than 2 °C (Jackson 2009). Furthermore, emission reductions expected from energy efficiency policies and carbon capture and storage systems are likely overestimated, mostly due to the energy penalty of some of those technologies, their scaling-up, non-CO_2 pollution, rebound effect on actual demand, transition costs, and vested interests (Arvesen et al. 2011; Sathre et al. 2011; see also Jacobson 2009). As Jackson (2009, p. 83)

[2] Implying about 455 ppm CO_2e, but with an actual warming impact similar to the raw CO_2 concentration value due to aerosol dimming and other offsetting factors (Hamilton 2010, p. 228).

points out, technological breakthroughs in energy generation, sequestration, or geo-engineering are not impossible, and could very well come from nanotechnology and synthetic biology (see also ETC Group 2004; Kunstler 2005; Hällström 2008). But they would need to overcome inherent risks and trade-offs of innovations, and kick in fast and large to generate the magnitude of decoupling needed between economic growth and the material-energy throughput. Yet no such technologies are more than unproven concepts, several years, and sometimes decades, away from testing, let alone safe commercialization. Furthermore, this says nothing of the political economy of their eventual deployment, notably the distribution of their benefits, costs, and risks (Moe 2010). Therefore, to suggest that, and act as if, socio-technological innovations will necessarily or even likely provide safe levels of decarbonization and sequestration, is dangerously delusive.

4.2 Assumptions About Adaptation

The above argument suggests that the ongoing failure of institutional reforms for adequate global environmental governance, and the unrealistic reliance on technological innovation, make it virtually impossible to avoid a gradual and possibly rapid decline of industrial agriculture outputs over the next few decades. This is true globally, and particularly for countries like Viet Nam that are highly vulnerable to climate change. Beyond such mythical decarbonization, the green economy discourse also rests on promises of robust adaptation, enabling to face whatever changes end up occurring. In the sphere of food production, this translates into climate-smart agriculture and a second wave of the *Green Revolution*.

Yet, the scope and depth of such adaptive measures is also unrealistic, while in fact industrial agriculture is proving to be inherently and ever more incapable of facing the challenges of multiple stressors, particularly that of climate change. First, the model constantly weakens its ecological foundation, as productivity is contingent upon continued and intensifying "biophysical override" (Weis 2010). It increases pressure on resources—notably water, land, and energy—and accelerates soil impoverishment, the inadvertent breeding of superweeds, the degradation of ecological services, and the spread of eco- and human-toxics. Those incompatibilities between industrial agriculture and the environment expose the former's abuses and systemic destabilization of the latter, and its increased fragility under environmental change (Blaikie and Brookfield 1987; Keil et al. 1998; Lipietz 2000; Adger et al. 2001; Bello 2004; Peet and Watts 2004; Bäckstrand and Lövbrand 2007, pp. 131–36; Forsyth 2003; Friedmann and McNair 2008; Vandermeer et al. 2009; McMichael 2009a; Moore 2010; Weis 2007, 2010; Foley et al. 2011). Ultimately, as Viet Namese ecosystems are disrupted, the natural resource base of livelihoods is necessarily weakened (Taylor 2007, pp. 11–12).

Second, the industrialization of agriculture has made farmers increasingly dependent on complex production and circulation processes, both within and outside agriculture. For example, up to a few decades ago, peasants of the Mekong

Delta used to *live with floods* rather than control them, profiting from alluvial fertilization. They grew varieties of rice that took longer to mature but better adapted to seasonal hydraulic variations (see notably Sneddon and Nguyen 2001; Tran et al. 2007, pp. 37–39). High dikes built across the Mekong Delta have reduced natural soil fertility for lack of sediments, leading to the increased use of agrochemicals, higher production costs, and increased water pollution. The opportunity for extra income and proteins that many poor households drew from fishing across flooded fields is also gone (Käkönen 2009, p. 208). The extent and magnitude of such landscape engineering required by industrial agriculture for irrigation, drainage, and salinity protection has in turn increased systemic sensitivity to environmental risks (for similar argument beyond Viet Nam, see Brooks et al. 2009, p. 751). Hydraulic systems are conceived within predictable ranges of variability, notably of water flow across river basins and weather patterns. Yet, such variability is increasingly unpredictable and reaches ever greater extremes, both as a result of resource overuse and of climate change. During the last decade, rainfall patterns have already changed in Viet Nam, with fewer precipitations in the north and more in the south (Institute of Strategy and Policy on Natural Resources and Environment 2009, p. xv). In addition, several new dams are being commissioned upstream on the Mekong—most significantly in China but also in Laos, Cambodia, and Thailand—which will affect the river's flow in yet uncertain ways but are likely to compete for depleting freshwater resources, and accelerate erosion, siltation, and salinization across the delta and within its aquifers (Vaidyanathan 2011).

Similarly, the process of modernization has prioritized accumulation through productivity at the expense of rural livelihoods and environmental sustainability, leading to the reclamation of much flood plains, wetlands, and shallow coastlines (see notably Adger 1999, p. 253; Lebel et al. 2009, p. 283), and increasing demand for both water control and use (Miller 2007, p. 199). The overexploitation of underground water for household use and economic activities has reached alarming levels, with research expecting the exhaustion of some Mekong Delta aquifers by 2013 if no urgent measures are taken (Nuber and Stolpe 2008; *VOV News* 2009). As sea level rises, the depletion of underground water and drainage of traditional flood plains makes flooding and saline penetration ever more frequent and consequential. All these factors will combine to make the country's extensive irrigation system less effective, and commensurately affect agricultural productivity (Sneddon and Nguyen 2001; Le 2007; *Viet NamNet Bridge* 2009; Käkönen 2009; Biggs et al. 2009, pp. 211–212; *Viet NamNet Bridge* 2010). This is particularly worrisome when considering that the Mekong Delta produces half of the country's rice and most of its surpluses and exports, but will also be one of the most severely affected region by freshwater shortages and sea-level rise. In a context of systemic dependency, the model of industrial agriculture is therefore nakedly exposing Viet Nam to a new vulnerability, weakening its resilience to external shocks, which could indeed void the successes achieved on food security over the last three decades.

Third, and maybe most importantly in the long term, industrial agriculture has affected the adaptability of Viet Namese farmers. The technologies of the Green

Revolution have led to monocultures of high-yielding varieties. Across Viet Nam's ecological zones, few modern and hybrid cultivars have replaced a large number of traditional ones that were well suited to their local climatic and biotic attributes, be it rain patterns, salinity, acidity, temperature, or pest prevalence. By constraining access to natural resources, reducing genetic diversity, knowledge, skills, and social networks, and now enclosing species within intellectual property fences, industrial agriculture has created technological and institutional path dependencies. This is foreclosing alternative options, some of which are becoming more than ever necessary for adapting to quickly evolving hydraulic and climatic conditions and to make the food production system more resilient (Adger 1999, p. 252–257; Perkins 2003; Young et al. 2006, p. 313; Taylor 2007, pp. 11–12; Food and Agriculture Organization of the United Nations 2008, pp. 44–45; Altieri and Koohafkan 2008; ETC Group 2009, pp. 8–14; Brooks et al. 2009, pp. 745–761; Havaligi 2009; Moore 2010, p. 399).

While aligned on the global discourse of ecological modernization, the Viet Namese response to the climate crisis therefore rests on unsubstantiated technological and institutional assumptions. On the one hand, the measures so far internationally agreed to mitigate GHG emissions fall well short of the magnitude and timing that the latest science indicates would have some chance of preserving climatic equilibrium. Furthermore, the potential of the technological solutions proposed to deliver such insufficient abatement, or to adapt to changes that cannot be avoided, is either overestimated or yet unproven, for want of research and operational deployment. On the other hand, Viet Nam's adaptation strategy is based on, and builds its expectations from, a paradigm threatened not only by multiple structural stressors but also by internal dynamics that render the model ever more precarious at three levels: it has weakened the ecological foundation of agriculture through *biophysical override*; it has increased dependence on complex production and circulation processes both within and outside agriculture; and it has locked the development of that agricultural model into path dependencies through technical, knowledge, and institutional homogenization that is now foreclosing alternatives, reducing options for resilience and adaptation.

5 Circular Metabolism

The underlying logic of industrial production and consumption, including modern agriculture in Viet Nam as discussed above, is essentially linear: an ever increasing amount of energy and material resources are mobilized at one end, creating both wealth and waste at the other (Murota 1998; Jones et al. 2011). Since the Earth is a finite system, except for a constant input of solar energy, the use of resources and the sinking of waste are bound to hit limits. This is now happening on numerous and widening fronts, including energy and freshwater scarcity, biodiversity losses, and climate change (Rockström et al. 2009a). To maintain functionality in a finite system, processes have to be circular, with all outputs being recycled into new inputs, with new energy (solar in the Earth's case) compensating for entropic loss. The argument of the previous sections underlines how much a new model is urgently needed, in Viet Nam as elsewhere, to address the inherent contradictions of industrial agriculture and emerging structural stressors.

For the past five decades, the realization of such limits and the conceiving and experimenting of alternatives have sprung and matured (including notably Bookchin 1962; Carson 1962; Georgescu-Roegen 1971; Meadows, Club of Rome, and Potomac Associates 1974; Blaikie and Brookfield 1987; Keil et al. 1998; Peet and Watts 2004; Foster et al. 2010). With respect to agriculture, limits and alternatives have also long been debated in both the scientific and policy literature (Weis 2007; McIntyre et al. 2009; McMichael 2009a; United Nations Conference on Trade and Development 2010; de Schutter 2010; Vía Campesina 2010; Moore 2010; Weis 2010; Woodhouse 2010; Hoffmann 2011; Foley et al. 2011). It has become clear that only by steering away from industrial processes will agriculture drastically reduce, and eventually reverse, its GHG emissions, and prevent the crossing of other planetary boundaries (Funes 2002; Borron 2006; Nelson et al. 2009a; Genetic Resources Action International 2009; Jones et al. 2011). Also, and crucially, the most robust of such alternatives underline the interaction of both technological and social dimensions, showing how ecologically sustainable solutions cannot be socially viable without a new organization of production, distribution, and consumption. Such coupling of social and technological dimensions implies the transformation of current linear practices toward a paradigm of circular metabolism, along with the restructuring of social relations that value such ecosystemic continuity and resilience well above the current model of individual-centric short-term maximized utility. This new socio-technical paradigm of agriculture is taking different forms and labels, including those of agroecology, permaculture, or ecological architecture, along with organizing principles of *food sovereignty* (Lee 2007; Pimbert 2008; Holt-Giménez 2009; Borras 2010; and Tran 2011 for the case of Viet Nam). The common thread between those models is that in order to effectively address the multifaceted crisis of agriculture while feeding a growing global population of 7–9 billion, four synchronous pathways of reforms must be advanced:

1. Agro-ecological production. To enable the circular metabolic objective of eliminating agrochemical inputs and reduce energy use, these will need to be substituted with higher labor intensity; smaller production scales; cooperative and reciprocity arrangements; and the de-globalization, localization, and shortening of commodity chains—transformed into proximity trading webs. It will also require the rebuilding of resilient agrobiodiversity, as well as a rich and dispersed knowledge in agroecological metabolism (Rosset 2003; Weis 2007; Friedmann and McNair 2008; Ploeg 2008; Altieri 2009; Weis 2010; Reardon and Perez 2010).
2. Socialized production. Such a transformation of production processes will necessarily confront the political economic resistance of interests vested in the current industrial agriculture and food system, from upstream agrochemical suppliers and farmers locked in technological and commercial path dependencies, to

downstream food processing industries, conglomerate retailers, and the array of financiers and marketeers all along that chain. To face this resistance to agroecology, and reclaim their long overdue control over production and accumulation, farmers will need to reappropriate their means of production, notably land, water, species, and knowledge, in a process that some have characterized as *repeasantization* (Ploeg 2007; Sevilla et al. 2009; Vanhaute 2011; see also Weis 2007; Borras 2008; Rosset 2009; McMichael 2009b; Rosset 2011; Torrez 2011).

3. Prioritized consumption. The linear character of industrial agriculture is at the root of its destructive impact, for its depletion of soils and biodiversity, its exhaustion of natural resources, and its environmental externalities—including climate change. The magnitude of energy and material imbalances also stems from the nature of goods such a system delivers, driven by market profitability rather than actual needs. While agroecology offers alternatives to the procedural dead-ends of industrial agriculture (*how* goods are produced), food sovereignty also undertakes to change *what* is produced. This is a necessary step in reducing the energy-material throughput and its pressure on ecosystems, while ensuring the most efficient response to societal needs. Most notably, this requires ending the production of biofuels and reducing wastes from food processing and commodification, but also demands a shift from energy-, water-, and grain-intensive commodities, such as meat and dairy products or all-season luxury fruits and vegetables, toward directly and locally consumable food of high nutritional quality and safety (ETC Group 2009; de Schutter 2009; Magdoff and Tokar 2009; McMichael 2009c; Stehfest et al. 2009; Foley et al. 2011).
4. Fair distribution. Close to a billion people are food insecure in the current system despite high levels of productivity and intensive resource use (Food and Agriculture Organization of the United Nations and World Food Programme 2010; Foley et al. 2011). This systemic failure is bound to increase in a world of ever more depleted resources and saturated environmental sinks. As discussed earlier, the rhetorical response of industrial agriculture to that crisis is to further intensify outputs through a renewed Green Revolution. In turn, food sovereignty argues that current production is sufficient, and even excessive, if it focuses on direct consumption and is more equally distributed. It reasserts the *social function* of food beyond commodity production and profit accumulation, as both livelihood and in fulfilling the universal right to food through self-consumption and community solidarity. Food sovereignty therefore establishes institutional redistribution mechanisms, such as in-kind grants to vulnerable groups, schools and hospitals, food-for-work, and affordable local markets (Schiavoni and Camacaro 2009; Wright 2009).

We have seen that the ecological modernization of the green economy rests on speculative technological and institutional assumptions. Those claimed solutions remain, however, to develop or to scale up, beyond what is realistic for the timing and magnitude needed. In contrast, agroecology and food sovereignty revert to long-proven agricultural practices, enhanced with newer but well-tested knowledge, and which provide sustainable use of resources, stronger biodiversity, and more resilience to environmental change, while delivering high levels of outputs (Altieri and Koohafkan 2008; Altieri 2009; McIntyre et al. 2009; de Schutter 2010; Vasilikiotis 2010; Vía Campesina 2010). This alternative model also rests on a sociopolitical organization of decentralized and democratized power over production, distribution, and access to food that has proven robust at times of crisis (for the example of Cuba, see Wright 2009; Reardon and Perez 2010; Machín et al. 2010; Rosset et al. 2011).

6 Conclusion

In Viet Nam over the past three decades, modernization has brought tremendous increases of food production and productivity, leading to significant surpluses and exports. There are, however, mounting threats, notably that of climate change, which could quickly void those gains and affect the country's food security. The response by Viet Namese authorities and most development actors has been to invest in further agricultural modernization, notably through expanded irrigation systems and the adoption of yet higher-yielding varieties and climate-proofed species. This reflects the modernization approach that Viet Nam has adopted for half a century, and the globally prevalent ecological modernization discourse of green capitalism. It makes a politically convenient act of faith in technological and institutional innovations, both for reducing GHG emissions to mitigate the magnitude of climate change, and for building the resilience of an industrial agriculture that would supposedly keep rising its productivity despite the challenging context. In fact, it is often argued that only industrial agriculture can possibly feed a growing world population (for example, Fedoroff et al. 2010).

Yet presuming of techno-institutional salvation not only fails to recognize the inherent limits of modernized agriculture but also obfuscates the instability of the model in the context of climate change, and how precarious it renders food security. The overall picture of Viet Nam's agriculture is by now one deeply and increasingly embedded in complex and global commodity chains requiring energy-intensive production, mechanical and chemical processing, packaging, transportation, storage, and refrigeration. It requires efficient norms, rules, and institutions of international trade, and a sustainable environment that provides the needed ecological services. Yet, energy supplies, agrochemicals, engineered processes such as irrigation and drainage, trade stability and, in particular, environmental change, are all areas of intensifying crisis. In the context of systemic dependency as described in this article, the embracing of industrial agriculture is now nakedly exposing Viet Nam to new sources of vulnerability, weakening its resilience to external shocks that could indeed blow away the deceivingly proud castle of cards it has built around agriculture and food security for the past 30 years.

Beyond the critique of that model, this article has argued that there is a compelling nexus between agroecology and food sovereignty that tackles the necessity of feeding a growing population with that of radically reducing agriculture's ecological footprint, while in fact rebuilding circular metabolic processes. That socio-technical paradigm is in this sense a far superior model to industrial agriculture, and there is a quickly growing body of literature discussing the theory and ongoing experiences of moving from the latter to the former in the broader context of a transition beyond industrial capitalism (see notably Bello 2004; Cheynet 2008; Victor 2008; Latouche 2009; Kempf 2009, pp. 115–34; Jackson 2009; Martínez Alier 2009). The next question is to see how such a paradigm can take hold in a country like Viet Nam, despite the creeping of interests now deeply vested in agro-industrial accumulation. Supporting the still very few agroecology and food sovereignty initiatives, and engaging in related policy debates, will be compelling and urgent contributions to that shift.

References

Adger WN (1998) Indicators of social and economic vulnerability to climate change in Viet Nam. CSERGE working paper GEC 98-02. The Centre for Social and Economic Research on the Global Environment (CSERGE), Norwich

Adger WN (1999) Social vulnerability to climate change and extremes in coastal Viet Nam. World Dev 27(2):249–269

Adger WN (2000) Social and ecological resilience: are they related? Prog Hum Geogr 24(3):347–364

Adger WN (2006) Vulnerability. Glob Environ Change 16(3):268–281

Adger WN, Barnett J (2009) Four reasons for concern about adaptation to climate change. Environ Plann A 41(12):2800–2805

Adger WN, Benjaminsen TA, Brown K, Svarstad H (2001) Advancing a political ecology of global environmental discourses. Dev Change 32(4):681–715

Altieri MA (2009) Agroecology, small farms, and food sovereignty. Monthly Rev Independent Socialist Mag 61(3):102–113

Altieri MA, Koohafkan P (2008) Enduring farms: climate change, smallholders and traditional farming communities 6. Third World Network (TWN), Penang

Anderson K, Bows A (2008) Reframing the climate change challenge in light of post-2000 emission trends. Philos Trans Royal Soc A Math Phys Eng Sci 366(1882):3863–3882

Arvesen A, Bright RM, Hertwich EG (2011) Considering only first-order effects? How simplifications lead to unrealistic technology optimism in climate change mitigation. Energ Policy 39(11):7448–7454

Asian Disaster Preparedness Center (ADPC) (2003) Climate change and development in Viet Nam: agriculture and adaptation for the Mekong delta region. ADPC and Deutsche Gesellschaft für Technische Zusammenarbeit (GTZ) GmbH

Bäckstrand K, Lövbrand E (2006) Planting trees to mitigate climate change: contested discourses of ecological modernization, green Governmentality and civic environmentalism. Glob Environ Polit 6(1):50–75

Bäckstrand K, Lövbrand E (2007) Climate governance beyond 2012: competing discourses of green governmentality, ecological modernization and civic environmentalism. In: Pettenger ME (ed) The social construction of climate change: power, knowledge, norms, discourses, Ashgate : Aldershot, Hampshire, England

Batchelor C, Schouten T, Smits S, Moriarty P, Butterworth J (2009) Climate change and WASH services delivery: is improved WASH governance the key to effective mitigation and adaptation? In: Perspective paper, prepared for the 5th world water forum. IRC International Water and Sanitation Centre, The Hague

Bates B, Kundzewicz ZW, Wu S, Palutikof J (2008) Climate change and water. Intergovernmental panel on climate change (IPCC) Technical Paper 6

Battisti DS, Naylor RL (2009) Historical warnings of future food insecurity with unprecedented seasonal heat. Science 323(5911):240–244

Bello W (2004) Deglobalization: ideas for a new world economy. New updated. Zed Books, London

Bhaduri A, Rahman MA (eds) (1982) Studies in rural participation. Oxford & IBH, New Delhi

Biggs D, Miller F, Thai Hoang C, Molle F (2009) The delta machine: water management in the Viet Namese mekong delta in historical and contemporary perspectives. In: Molle F, Foran T, Käkönen M (eds) Contested waterscapes in the mekong region: hydropower, livelihoods and governance, Earthscan, London

Binh D, Lam V, Nam N, Nam D (2010) Thieu nuoc, dien phap phu [water shortage, unstable supply of electricity]. Tuoi Tre Online, 10 Sep 2010

Blaikie PM, Brookfield HC (eds) (1987) Land degradation and society. Methuen, London; New York

Bookchin M (1962) Our synthetic environment, 1st edn. Knopf, New York

Borras SM (2008) La Vía Campesina and its global campaign for agrarian reform. J Agrarian Change 8(2–3):258–289

Borras SM (2010) The politics of transnational agrarian movements. Dev Change 41(5):771–803

Borron S (2006) Building Resilience for an unpredictable future: how organic agriculture can help farmers adapt to climate change. Food and Agriculture Organization of the United Nations (FAO), Rome

Brooks N, Grist N, Brown K (2009) Development futures in the context of climate change: challenging the present and learning from the past. Dev Policy Rev 27(6):741–765

Carew-Reid J (2008) Rapid assessment of the extent and impact of sea level rise in Viet Nam. International Centre for Environmental Management (ICEM)

Carson R (1962) Silent spring. Houghton Mifflin, Boston

Chaudhry P, Ruysschaert, G (2007) Climate change and human development in Viet. United Nations Development Programme (UNDP) Human Development Report Office Occasional Paper 2007/42

Cheynet V (2008) Le choc de la décroissance, histoire immédiate. Éditions du Seuil, Paris

Chorev N, Babb S (2009) The crisis of neoliberalism and the future of international institutions: a comparison of the IMF and the WTO. Theory Soc 38(5):459–484

Cline WR (2007) Global warming and agriculture: impact estimates by country. Center for Global Development, Washington

Dasgupta S, Laplante B, Meisner G, Wheeler D, Yan J (2009) The impact of sea level rise on developing countries: a comparative analysis. Climatic Change 93(3):379–388

Eakin H, Luers AL (2006) Assessing the vulnerability of social-environmental systems. Ann Rev Environ Resour 31(1):365–394

Ericksen PJ (2008) What is the vulnerability of a food system to global environmental change? Ecol Soc 13(2)

Ericksen P, Thornton P, Notenbaert A, Cramer L, Jones P, Herrero M (2011) Mapping hotspots of climate change and food insecurity in the global tropics. CGIAR research program on climate change, agriculture and food security (CCAFS)

ETC Group (2004) Down on the farm: the impact of nano-scale technologies on food and agriculture. ETC Group, Ottawa

ETC Group (2009) Who will feed us? Questions for food/climate crises negotiators in Rome and Copenhagen. ETC Group, Ottawa

Fedoroff NV, Battisti DS, Beachy RN, Cooper PJM, Fischhoff DA, Hodges CN, Knauf VC, Lobell D, Mazur BJ, Molden D, Reynolds MP, Ronald PC, Rosegrant MW, Sanchez PA, Vonshak A, Zhu JK (2010) Radically rethinking agriculture for the 21st century. Science 327(5967):833–834

Fforde A, Sénèque S (1994) The economy and the countryside in Viet Nam: the relevance of rural development policies. Research School of Pacific and Asian Studies, Canberra

Foley JA, Ramankutty N, Brauman KA, Cassidy ES, Gerber JS, Johnston M, Mueller ND, O'Connell C, Ray DK, West PC, Balzer C, Bennett EM, Carpenter SR, Hill J, Monfreda C, Polasky S, Rockstrom J, Sheehan J, Siebert S, Tilman D, Zaks DPM (2011) Solutions for a cultivated planet. Nature 478(7369):337–342

Food and Agriculture Organization of the United Nations (FAO) (2008) Climate change and food security: a framework document. FAO, Rome

Food and Agriculture Organization of the United Nations (FAO) and World Food Programme (WFP) (2010) The state of food insecurity in the world: addressing food insecurity in protracted crises. FAO and WFP, Rome

Forsyth T (2003) Critical political ecology: the politics of environmental science. Routledge, New York

Fortier F (2010) Taking a climate chance: a procedural critique of Viet Nam's Climate Change Strategy. Asia Pac Viewpoint 51(3):229–247

Foster EA (2008) Sustainable development policy in Britain: shaping conduct through global governmentality. Br Politics 3(4):535–555

Foster JB, Clark B, York R (2010) The ecological rift: capitalism's war on the earth. Monthly Review Press, New York

Friedmann H, McNair A (2008) Whose rules rule? Contested projects to certify 'local production for distant consumers. J Agrarian Change 8(2–3):408–434

Funes F (2002). Sustainable agriculture and resistance: Transforming food production in Cuba. Food First Books, Oakland

Füssel H-M (2007) Vulnerability: a generally applicable conceptual framework for climate change research. Global Environ Change 17(2):155–167

General Statistics Office of Viet Nam (GSO) (2009a) Key indicators on national accounts. GSO, Hanoi

General Statistics Office of Viet Nam (GSO) (2009b) Production of paddy by province. GSO, Hanoi

Genetic Resources Action International (GRAIN) (2008) Viet Nam: The high stakes of hybrid rice for farmers. GRAIN, Barcelona

Genetic Resources Action International (GRAIN) (2009) Climate crisis special issue. GRAIN, Barcelona

Georgescu-Roegen N (1971) The entropy law and the economic process. Harvard University Press, Cambridge

Gills BK (2010) Going South: capitalist crisis, systemic crisis, civilisational crisis. Third World Q 31(2):169–184

Government of Viet Nam (2008) Quyết định phê duyệt chương trình mục tiêu quốc gia ứng phó với biến đổi khí hậu [Approval of the National Target Program in Response to Climate Change]. Edited by Thủ Tướng Chính Phủ (Prime Minister Office), Hanoi

Government of Viet Nam (2009) Nghi quyet so 63/NQ-CP ngay 23 thang 12 nam 2009 cua chinh phu ve dam bao an ninh luong thuc quoc gia [Government Resolution No. 63/NQ-CP of 23 December 2009 on Ensuring National Food Security], Hanoi

Granich S, Kelly PM, Nguyen HN (1993) Global warming and Viet Nam. Stockholm Environment Institute, Stockholm

Greenfield G (2004) Viet Nam and the WORLD COFFEE CRISIS: Local coffee riots in a global context. http://www.probeinternational.org/coffee/Viet Nam-and-world-coffee-crisis-local-coffee-riots-global-context. Accessed 22 Sep 2010

Vinh Ha (1997) Nong nghiep Viet Nam trong buoc chuyen sang kinh te thi truong. Social Sciences Publishing House, Hanoi

Hällström N (2008) What next? Climate change technology and development. Development 51(3):375

Hamilton C (2010) Requiem for a species: why we resist the truth about climate change. Earthscan, London; Washington

Hansen J (2005) A slippery slope: how much global warming constitutes' dangerous anthropogenic interference? Climatic Change 68(3):269–279

Hansen J, Sato M, Kharecha P, Beerling D, Berner R, Masson-Delmotte V, Pagani M, Raymo M, Royer DL, Zachos JC (2008) Target atmospheric CO2: where should humanity aim? Open Atmos Sci J 2:217–231

Havaligi N (2009) Role of agrobiodiversity and indigenous knowledge conservation in food and global health security in the climate change scenario. In: Climate 2009. Virtual Conference

Hoffmann U (2011) Assuring food security in developing countries under the challenges of climate change: key trade and development issues of a fundamental transformation of agriculture. United Nations conference on trade and development (UNCTAD) Discussion Paper 201

Holt-Giménez E (2009) From food crisis to food sovereignty. Monthly Rev Independent Socialist Mag 61(3):142–156

Hong Van (2007) Gia phan bon tang manh (The Price of Fertilisers Increased Sharply). Thoi bao kinh te Sai Gon online

IEA (2010) World Energy Outlook 2010—Executive Summary. International Energy Agency, Paris:14

Institute of Strategy and Policy on Natural Resources and Environment (ISPONRE) (2009) Viet Nam assessment report on climate change. ISPONRE

Intergovernmental Panel on Climate Change (IPCC) (2007) Climate change 2007: impacts, adaptation and vulnerability: contribution of working group II to the fourth assessment report of the intergovernmental panel on climate change. Edited by the IPCC working group II. Cambridge University Press, Cambridge

International Centre for Environmental Management (ICEM) (2009) Ho Chi Minh city adaptation to climate change. ICEM

International Monetary Fund (IMF) (2010) World economy outlook database

International Rice Research Institute (IRRI) (2007) Important rice production system under pressure. ScienceDaily, 10 Oct

International Rice Research Institute (IRRI) (2009) IRRI world rice statistics (WRS). http://beta.irri.org/solutions/index.php?option=com_content&task=view&id=250

Jackson T (2009) Prosperity without growth: economics for a finite planet. 1st edn. Earthscan, London

Jacobson MZ (2009) Review of solutions to global warming, air pollution, and energy security. Energy Environ Sci 2(2):148–173

Jones A, Pimbert M, Jiggins J (2011) Virtuous circles: values, systems and sustainability, reclaiming diversity and citizenship. International Institute for Environment and Development, London

Käkönen M (2009) Mekong delta at the crossroads: more control or adaptation? AMBIO J Human Environ 37(3):205–212

Keil R, Bell DVJ, Penz P, Fawcett L (eds) (1998) Political ecology: global and local. innis centenary series. Routledge, New York

Kelly PM, Adger WN (2000) Theory and practice in assessing vulnerability to climate change and facilitating adaptation. Climatic Change 47(4):325–352

Kempf H (2009) Pour sauver la planète, sortez du capitalisme, L'histoire immédiate. Éditions du seuil, Paris

Kerkvliet BJT (1995) Village-state relations in Viet Nam: the effect of everyday politics on decollectivization. J Asian Stud 54(2):396–418

KOF (2007) KOF index of globalisation. KOF Konjunkturforschungstelle

Kolko G (1997) Viet Nam: anatomy of a peace. Routledge, New York

Kunstler JH (2005) The long emergency: surviving the converging catastrophes of the twenty-first century. Atlantic Monthly Press, New York

Latouche S (2009) Farewell to growth. Polity, Cambridge

Le SL (2007) Stink over Viet Namese food exports. Asia Times Online. 12 Dec

Lebel L, Sinh BT, Garden P, Seng S, Tuan LE, Truc DV (2009) The promise of flood protection: dikes and dams, drains and diversions. In: Molle F, Foran T, Käkönen M (eds) Contested waterscapes in the mekong region: hydropower, livelihoods and governance, Earthscan, London

Lee R (2007) Food security and food sovereignty. Centre for Rural Economy, University of Newcastle Upon Tyne

Li M (2007) "Peak Oil, the Rise of China and India, and the Global Energy Crisis". J Contemp Asia 37(4):449–471

Lipietz A (2000) Political ecology and the future of Marxism. Capitalism Nat Socialism 11(1):69–85

Luke TW (1996) Generating green governmentality: a cultural critique of environmental studies as a power/knowledge formation. Department of Political Science, Virginia Polytechnic Institute and State University Blacksburg, Blacksburg

Machín S, Braulio, Jaime AMR, Ávila Lozano DR, Rosset PM (2010) Revolución agroecologica: El movimiento de campesino a campesino de la ANAP en Cuba. La Vía Campesina

Magdoff F, Tokar B (2009) Agriculture and food in crisis. Monthly Rev Independent Socialist Mag 61(3):1–16

Marquis JP (2000) The other warriors: American social science and nation building in Viet Nam. Diplomatic Hist 24(1):79

Martínez Alier J (2009) Socially sustainable economic de-growth. Dev Change 40(6):1099–1119

Martínez Alier J (2011) The EROI of agriculture and its use by the via Campesina. J Peasant Stud 38(1):145–160

Max-Neef M (2010) The world on a collision course and the need for a new economy. AMBIO J Human Environ 39(3):200–210

McIntyre BD, Herren HR, Wakhungu J, Watson RT (eds) (2009) Agriculture at a crossroad: international assessment of agricultural knowledge, science and technology for development (IAASTD)—global report. Island Press, Washington

McMichael P (2009a) A food regime analysis of the world food crisis. Agric Hum Values 26(4):281–295

McMichael P (2009b) A food regime genealogy. J Peasant Stud 36(1):139–169

McMichael P (2009c) Banking on agriculture: a review of the world development report 2008. J Agrarian Change 9(2):235–246

Meadows DH (1974) Club of Rome, Potomac associates. The limits to growth: a report for the club of Rome's project on the predicament of mankind. 2nd edn. Universe Books, New York

Miller F (2007) Seeing 'water blindness': water control in agricultural intensification and environmental change in Mekong Delta, Viet Nam. In: Connell J, Waddell E (eds) Environment, development and change in rural Asia-Pacific: between local and global. Routledge, New York

Ministry of Agriculture and Rural Development (MARD) (2008) Action Plan framework for adaptation to climate change in the agriculture and rural development sector period 2008–2020. Hanoi

Ministry of Natural Resources and Environment of Viet Nam (MONRE) (2003) Viet Nam initial national communication under the united nations framework convention on climate change. Hanoi

Ministry of Natural Resources and Environment of Viet Nam (MONRE) (2008) Viet Nam assessment report on climate change (VARCC). Hanoi

Moe E (2010) Energy, industry and politics: energy, vested interests, and long-term economic growth and development. Energy 35(4):1730–1740

Moore JW (2010) The End of the Road? Agricultural revolutions in the capitalist world-ecology, 1450–2010. J Agrarian Change 10(3):389–413

Murota T (1998) Material cycle and sustainable economy. In: Keil R, Bell DVJ, Penz P, Fawcett L (eds) Political ecology: global and local. Routledge, London

Nellemann C, MacDevette M, Manders T, Eickhout B, Svihus B, Prins AG, and Kaltenborn BP (eds) (2009) The Environmental food crisis—the environment's role in averting future food crises. A UNEP rapid response assessment united nations environment programme (UNDP), Arendal, Norway

Nelson DR, Adger WN, Brown K (2007) Adaptation to environmental change: contributions of a resilience framework. Ann Rev Environ Res 32(1):395–419

Nelson E, Scott S, Cukier J, Galán A (2009a) Institutionalizing agroecology: successes and challenges in Cuba. Agric Hum Values 26(3):233–243
Nelson GC, Rosegrant MW, Koo J, Robertson R, Sulser T, Zhu T, Ringler C, Msangi S, Palazzo A, Batka M, Magalhaes M, Valmonte-Santos R, Ewing M, Lee D (2009b) Climate change: impact on agriculture and costs of adaptation. International Food Policy Research Institute (IFPRI)
Ngo VL (1984) Agrarian differentiation in the southern region of Viet Nam. J Contemp Asia 14(3):283–305
Ngo VL (1993) Reform and rural development: impact on class, sectoral, and regional inequalities. In: Turley WS, Selden M (eds) Reinventing Viet Namese socialism: Doi Moi in comparative perspective. Westview Press, Boulder
Nguyen HN, Trung VK, Nguyen XN (2007) Flooding in Mekong River Delta, Viet Nam. United Nations Development Programme (UNDP) Occasional Papers 2007/53: 23
Nguyen VV (2011) Climate change and agricultural production in Viet Nam. In: Leal Filho W (ed) Climate change management: economic, social and political elements of climate change, Springer, Berlin
Nuber T, Stolpe H (2008) Challenges of the groundwater management in Can Tho City, Viet Nam. In BGR- symposium, sanitation and groundwater protection. Hannover
O'Brien KL, Leichenko RM (2000) Double exposure: assessing the impacts of climate change within the context of economic globalization. Global Environ Change 10(3):221–232
O'Brien KL, Leichenko R, Kelkar U, Venema H, Aandahl G, Tompkins H, Javed A, Bhadwal S, Barg S, Nygaard L, West J (2004) Mapping vulnerability to multiple stressors: climate change and globalization in India. Global Environ Change Part A 14(4):303–313
OXFAM International (2008) Viet Nam: climate change, adaptation and poor people. Oxfam International, Hanoi
Pacala S, Socolow R (2004) Stabilization wedges: solving the climate problem for the next 50 years with current technologies. Science 305(5686):968–972
Panitch L, Gindin S (2009) The current crisis: a socialist perspective. Stud Political Econ 83:7–31
Pearce KC (2001) Rostow, Kennedy, and the rhetoric of foreign aid, rhetoric and public affairs series. Michigan State University Press, East Lansing
Peet R, Watts M (2004) Liberation ecologies: environment, development, social movements, 2nd edn. Routledge, New York
Perkins R (2003) Technological "lock-in". Internet Encyclopaedia of Ecological Economics, International Society for Ecological Economics. Online paper http://www.docin.com/p-423147640.html, retrieved 1 Sep 2012
Philipps P (2007) Peak oil and the future of agriculture: has Cuba Shown the way? In: Black Point NS (ED) Energy security and climate change: a Canadian primer. Fernwood Pub
Pimbert M (2008) Towards food sovereignty: reclaiming autonomous food systems. IIED
Pincus J, Sender J (2008) Quantifying poverty in Viet Nam: who counts? J Viet Namese Stud 3(1):108–150
Pingali PL, Nguyen TK, Gerpacio RV, Xuan TV (1997) Prospects for sustaining Viet Nam's reacquired rice exporter status. Food Policy 22(4):345–358
van der Ploeg JD (2007) The third agrarian crisis and the re-emergence of processes of repeasantization. Rivista di Economia Agraria 62(3):325–332
van der Ploeg JD (2008) The new peasantries: struggles for autonomy and sustainability in an era of empire and globalization. Earthscan, London
Radice H (2009) Neoliberalism in crisis? Money and the state in contemporary capitalism. Spectr J Global Stud 1(2). http://www.herramienta.com.ar/print/foro-capitalismo-en-trance/neoliberalism-crisis-money-and-state-contemporary-capitalism, retrieved 3 Jan 2013
Reardon JAS, Perez RA (2010) Agroecology and the development of indicators of food sovereignty in Cuban food systems. J Sustainable Agric 34(8):907–922
Richardson K (2009) Climate change: global risks, challenges and decisions (synthesis report). Paper read at climate change: global risks, challenges and decisions (synthesis report), Copenhagen, 10–12 March

Rockström J, Steffen W, Noone K, Persson A, Chapin FS, Lambin EF, Lenton TM, Scheffer M, Folke C, Schellnhuber HJ, Nykvist B, de Wit CA, Hughes T, van der Leeuw S, Rodhe H, Sorlin S, Snyder PK, Costanza R, Svedin U, Falkenmark M, Karlberg L, Corell RW, Fabry VJ, Hansen J, Walker B, Liverman D, Richardson K, Crutzen P, Foley JA (2009a) A safe operating space for humanity. Nature 461(7263):472–475

Rockström J, Steffen W, Noone K, Persson A, Chapin FS, Lambin EF, Lenton TM, Scheffer M, Folke C, Schellnhuber HJ, Nykvist B, de Wit CA, Hughes T, van der Leeuw S, Rodhe H, Sorlin S, Snyder PK, Costanza R, Svedin U, Falkenmark M, Karlberg L, Corell RW, Fabry VJ, Hansen J, Walker B, Liverman D, Richardson K, Crutzen P, Foley JA (2009b) Planetary boundaries: exploring the safe operating space for humanity. Ecol Soc 14(2). http://www.ecology and society.org/vol14/iss2/art32/, retrieved 3 Jan 2013

Rogelj J, Hare W, Lowe J, van Vuuren DP, Riahi K, Matthews B, Hanaoka T, Jiang K, and Meinshausen M (2011) Emission pathways consistent with a 2C global temperature limit. Nature Climate Change. Advance online publication

Rosegrant MW, Ewing M, Yohe G, Burton I, Huq S, Valmonte-Santos R (2008) Climate change and agriculture, threats and opportunities. Deutsche Gesellschaft für Technische Zusammenarbeit (GTZ) GmbH

Rosset P (2003) Food sovereignty: global rallying cry of farmer movements. Food First Backgrounder, Fall

Rosset P (2009) Fixing our global food system. Monthly Rev Independent Socialist Mag 61(3):114–128

Rosset PM (2011) Food sovereignty and alternative paradigms to confront land grabbing and the food and climate crises. Development 54(1):21–30

Rosset PM, Braulio MS, Jaime AMR, Lozano DRA (2011) The Campesino-to-Campesino agroecology movement of ANAP in Cuba: social process methodology in the construction of sustainable peasant agriculture and food sovereignty. J Peasant Stud 38(1):161–191

Sathre R, Chester M, Cain J, Masanet E (2011) A framework for environmental assessment of CO2 capture and storage systems. Energy (0)

Schiavoni C, Camacaro W (2009) The Venezuelan effort to build a new food and agriculture system. Monthly Rev Independent Socialist Mag 61(3):129–141

Schmidhuber J, Tubiello FN (2007) Global food security under climate change. Proc Natl Acad Sci 104(50):19703–19708

Schutter OD (2009) The Meatification of Diets and Global Food Security, UN Special Rapporteur on the right to food, Brussels http://www.europarl.europa.eu/climatechange/doc/speeche_Mr_de_schutter.pdf, retrieved 1 Jan 2012

Schutter OD (2010) The Right to Food: Report submitted by the Special Rapporteur on the right to food, Olivier De Schutter, United Nations General Assembly, New York, 11 August.http://www.srfood.org/images/stories/pdf/officialreports/20101021_access-to-land-report_en.pdf, retrieved 1 Jan 2012

Sevilla G, Eduardo Montiel MS (2009) Del desarrollo rural a la agroecologia. Hacia un cambio de pardigma [From Rural Development to Agroecology: Toward a Change of Paradigm]. Documentación social 155:23–39

Smit B, Wandel J (2006) Adaptation, adaptive capacity and vulnerability. Global Environ Change 16(3):282–292

Smith JB, Schneider SH, Oppenheimer M, Yohe GW, Hare W, Mastrandrea MD, Patwardhan A, Burton I, Corfee-Morlot J, Magadza ZHD, Füssel HS, Pittock AB, Rahman A, Suarez A, van Ypersele JP (2009) Assessing dangerous climate change through an update of the intergovernmental panel on climate change (IPCC) 'reasons for concern'. Proc Natl Acad Sci 106(11):4133–4137

Sneddon C, Nguyen TB (2001) Politics, ecology and water: the Mekong Delta and development of the lower Mekong Basin. In: Adger WN, Kelly PM, Nguyen HN (eds) Living with environmental change: social vulnerability, adaptation and resilience in Viet Nam, Routledge, London

Stehfest E, Bouwman L, van Vuuren D, den Elzen M, Eickhout B, Kabat P (2009) Climate benefits of changing diet. Clim Change 95(1):83–102

Taylor P (2007) Poor policies, wealthy peasants: alternative trajectories of rural development in Viet Nam. J Viet Namese Stud 2(2):3–56
Torrez F (2011) La Via Campesina: peasant-led agrarian reform and food sovereignty. Development 54(1):49–54
Tran TB, Sinh BT, Miller F (eds) (2007) Challenges to sustainable development in the Mekong Delta: regional and national policy issues and research needs. Bangkok, Sumernet
Tran TTT (2009a) State-society relations and the diversity of peasant resistance in Viet Nam. In: Caouette D, Turner S (eds) Agrarian Angst and rural resistance in contemporary Southeast Asia. Routledge, New York
Tran TTT (2010) Social differentiation revisited: a study of rural changes and peasant strategies in Viet Nam. Asia Pac Viewpoint 51(1):17–35
Tran T (2009) Climate change in Viet Nam and response. In: 7th Fig (international federation of surveyors) regional conference: spatial data serving people: land governance and the environment—building the capacity. Viet Nam Institute of Meteorology, Hydrology and Environment: Hanoi
Tran TTT (2011) Food Security vs. Food Sovereignty: Choice of Concept, Choice of Policies, and Choice of Classes in Viet Nam's Post-Reform. "Kasarinlan", Philippine J Third World Stud 26(1–2):66–88
Turral H, Burke J, Faurès JM (2011) Climate change, water and food security. United Nations Food and Agriculture Organization FAO Water Reports 36
UK Government Office for Science (2011) Foresight—the future of food and farming: challenges and choices for global sustainability. The Government Office for Science
United Nations Conference on Trade and Development (UNCTAD) (2010) Agriculture at the crossroads: guaranteeing food security in a changing global climate. UNCTAD Policy Briefs 18
United Nations Environment Programme (UNEP) (2011) Bridging the emissions gap. United Nations Environment Programme
United Nations Convention on Climate Change (UNFCCC) (2006) Technologies for adaptation to climate change. UNFCCC
Vaidyanathan G (2011) Dam controversy: remaking the Mekong. Nature 478(7369):305–307
Vandermeer J, Smith G, Perfecto I, Quintero E (2009) Effects of industrial agriculture on global warming and the potential of small-scale agroecological techniques to reverse those effects. La Via Campesina
Vanhaute E (2011) From famine to food crisis: what history can teach us about local and global subsistence crises. J Peasant Stud 38(1):47–65
Vasilikiotis C (2010) Can organic farming 'feed the world'? In agroecology and sustainable food systems. Food First, Institute for Food and Development Policy, Oakland
Vía Campesina L (2010) Sustainable peasant and family farm agriculture can feed the world. La Vía Campesina
Victor PA (2008) Managing Without Growth: Slower by Design, Not Disaster, Advances in Ecological Economics. Edward Elgar, Cheltenham
Viet Nam Business News (2010) Fertiliser prices rising fast. Viet Nam Business News, 20 April
Viet Nam News (2011) Climate change script reflects harsh realities. 8 Sep
Viet Nam Peasant Association (2010) Cat dien nong thon trien mien: San xuat lay lat, nguoi song quay quat" [Continous Blackout in the Countryside: Production Dropping, Humans Suffering]. Nien giam Nong nghiep—Thuc pham, 15 May
Viet NamNet Bridge (2009) UN Says China Dams Threaten Water Supplies to Mekong Delta Farmers. May 25
Viet NamNet Bridge (2010) Activists argue that dams will kill the mighty Mekong. 9 Feb
Vogel C, O'Brien K (2004) Vulnerability and global environmental change: rhetoric and reality. AVISO (13)
VOV News (2009) Nước ngầm tại Đồng bằng sông Cửu Long tụt giảm mạnh [Underground Water in the Mekong Delta Has Depleted Significantly]. 19 April
Weis A (2007) The global food economy: the battle for the future of farming. Zed Books, Halifax, Fernwood Pub, London

Weis A (2010) The accelerating biophysical contradictions of industrial capitalist agriculture. J Agrarian Change 10(3):315–341
Wiegersma N (1988) Viet Nam: peasant land, peasant revolution: patriarchy and collectivity in the rural economy. MacMillan Press, Hong Kong
Woodhouse P (2010) beyond industrial agriculture? Some questions about farm size, productivity and sustainability. J Agrarian Change 10(3):437–453
World Bank (2010) World development report 2010: development and climate change. The World Bank, Washington
Wright J (2009) Sustainable agriculture and food security in an era of oil scarcity: lessons from Cuba. Earthscan, London
Young KB., Wailes EJ, Cramer GL, Nguyen TK (2002) Viet Nam's rice economy: developments and prospects. University of Arkansas Research Report 968
Young OR, Berkhout F, Gallopin GC, Janssen MA, Ostrom E, van der Leeuw S (2006) The globalization of socio-ecological systems: an agenda for scientific research. Global Environ Change 16(3):304–316
Yu B, Zhu T, Breisinger C, Nguyen MH (2010) Impacts of climate change on agriculture and policy options for adaptation: the case of Viet Nam. IFPRI discussion paper 01015. IFPRI: Washington

MIX
Papier aus verantwortungsvollen Quellen
Paper from responsible sources
FSC® C105338

If you have any concerns about our products,
you can contact us on
ProductSafety@springernature.com

In case Publisher is established outside the EU,
the EU authorized representative is:
Springer Nature Customer Service Center GmbH
Europaplatz 3, 69115 Heidelberg, Germany

Printed by Libri Plureos GmbH
in Hamburg, Germany